The
Huckleberry Book

by 'Asta Bowen
with illustrations by Karen Leigh Noice

MONTANA
M A G A Z I N E

American Geographic Publishing
Helena, Montana

William A. Cordingley, Chairman
Rick Graetz, Publisher
Mark Thompson, Director of Publications
Carolyn Cunningham, Editor, Montana Magazine
Barbara Fifer, Assistant Book Editor

Dedicated to all
friends of the huckleberry everywhere

and to Susie
Who Would Have Loved Them
Too

Library of Congress Cataloging-in-Publication Data

Bowen, 'Asta.
   The huckleberry book.

   1. Huckleberries. 2. Huckleberries--West (U.S.) 3. Cookery (Huckleberries) I. Title.
SB 386.H83B69 1988   583'.62   88-16798
ISBN 0-938314-46-7 (pbk.)

ISBN 0-938314-46-7

© 1988 American Geographic Publishing, Box 5630,
Helena, Montana 59604. (406) 443-2842. All rights reserved.

Text © 1988 'Asta Bowen

Printed in U.S.A.

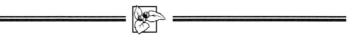

Acknowledgments

To everyone who came forward with stories, opinions, recipes, research, and enthusiasm, may this book honor the huckleberry as you do.

The science of huckleberries owes much to Nellie Stark, Pete Stickney and Don Minore. The history of huckleberries has been enhanced by the thorough curiosity of Francis Pettit. The business of huckleberries grows daily, thanks to the efforts of Eva Gates Preserves, Rena's Kitchen, Claude Sterner, Don Dwyer, Helen Bundrock, and other entrepreneurs I have not had the privilege to meet. To these, and to the berriers who live and tell their stories, this book owes its existence.

Special thanks to all who shared their huckleberry secrets, and regards to those who kept a few in reserve.

Quotations

Page 10, "Huckleberries and the Aurora Borealis," *Kim Williams' Cookbook and Commentary*, © 1983 by Kim Williams. Reprinted by permission of the publisher, HPBooks, a division of Price Stearns Sloan, Inc.

Page 11, "If you want to call a certain berry huckleberry...," *Eating Wild Plants*, Kim Williams, Mountain Press Publishing Company

Page 26, "...as we had nothing but a little flour...," from *The Journals of Lewis and Clark*, by Bernard DeVoto. © 1953 by Bernard DeVoto. © renewed 1981 by Avis DeVoto. Reprinted by permission of Houghton Mifflin Company.

Page 29, "When I was about nine years old...," *Flathead and Kootenai*, Olga Johnson, Arthur Clark Company

Page 38, "Say Huckleberry, But Keep Your Lucky Star Secret," *The Makeup of Ice*, by Paul Zarzyski, © 1984 the University of Georgia Press, reprinted by permission of the University of Georgia Press.

Page 58, "Sometimes, having had a surfeit...," *Walden*, Henry David Thoreau, Doubleday

Recipes

Huckleberry Dumplings from *Berry Good Cookbook*, Trout Creek Improvement Association

Huckleberry Jelly from *Butte's Heritage Cookbook*, Butte Bicentennial Association

Huckleberry Tea from *Canyon Cookery*, Bridger Canyon Women's Club

Rocky Mountain Huckleberry Cake from *Eating Wild Plants*, Kim Williams, Mountain Press Publishing Company

Kim Williams' Husband's Montana Huckleberry Pie from *Kim Williams' Cookbook and Commentary*, © 1983 by Kim Williams. Reprinted by permission of the publisher, HPBooks, a division of Price Stearns Sloan, Inc.

Huckleberry Muffins, Huckleberry Fritters, Pemmican from *Stalking the Wild Asparagus*, by Euell Gibbons, © 1962 by Euell Gibbons, published by David McKay Co.

Sour Cream Huckleberry Pancakes, Huckleberry Bran Muffins, Huckleberry Upside Down Cake, Carrie's Cake, Huckleberry Rice Salad from *Swan Lake Huckleberry Cookbook*, Swan Lake Women's Club

Frozen Huckleberries from *Trout Creek Huckleberry Recipes*, Trout Creek Improvement Association

Also: Ted Anders, R.C. Beall, Mildred Chaffin, Helen & Hilda Doten, Jan Dunbar, Allison Easterling, Tom Elgan, Faith Hodges, Barbara Hoffman, Gay Moddrell, Heather Mull, Betty Schopp, Ted Schwinden, Helen Turner

Say huckleberry,
but keep your lucky star secret

—Paul Zarzyski

Huckleberries Are

Into the Jewel

The Jewel Basin in Montana's Swan Range is a
compact wilderness area known for its pristine lakes,
subalpine wildflowers, bear, deer, elk and—as it happens—
huckleberries.

Seek the source,
the soul of the prism,
where what is priceless
is most plentiful.

Enter emerald.
Carve a trail
of topaz up mountains
studded with garnet,
to meadows
littered with amethyst.
Stumble among sapphire,
one immense and perfect sky
you carry like a crown.

Live and die
among huckleberries big as diamonds
you always have and always hunger for,
becoming yourself
precious and plentiful.

You can take the huckleberry out of the mountains
but you can't take the mountains out of the huckleberry.

Huckleberries are wrapped in secrecy and hidden in the wilderness, and only come out every other year. Or, huckleberries are everywhere in abundance always, and anyone can find them whenever they please.

Huckleberries are sweet. Huckleberries are sour. Huckleberries are woman's work, or a job for a man. Huckleberries are bigger in the shade, or sometimes bigger in the sun; huckleberries are easier to pick with rakes, but should only be picked by hand.

Huckleberries are really blueberries...no! nothing like blueberries. Huckleberries are worth risking your life for—or one good reason for living.

What is a huckleberry? Is it a time of year or a kind of freshness or a view from a mountain, or all of this sized and stored in a roundness and a sweetness and a purpleness we can name and touch and eat?

The huckleberry is wildness in your hand, a mountain summer you can savor in the dark of a winter morning. Imagine eating wildness! Imagine it storming your veins all day, coursing your heart like a western river, lining your bones with what gives the grizzly its grunt. You could be anywhere—Chicago, Bombay, the classroom—and each step would be on a fresh trail, each breath one of sun. You would walk with that lilt, the lilt of legs that know the way up a mountain. Imagine eating wildness.

What is wild has its secrets, and this is true of huckleberries. What is wild does not announce itself on television or hire important people to give it identity; what is wild has identity to spare, and needs no introduction. What is wild goes on with or without our acknowledgment, firm and perfect in all its variations. The huckleberry grows all day long on hillsides spread across mountaintops anyone can see: no guards, no gates, no passwords. Secret? Scarcely. Private? Perhaps.

Yes, the huckleberry leads a private existence, a life of quiet dignity far removed from the whir of civilization and science. Its habitat just happens to be all but inaccessible to human beings, a home well-chosen for one who keeps silent counsel with bears and wildfire and deep mountain snows.

And so the inner workings by which the huckleberry has persevered and prospered remain mostly unknown. Why does a stand flourish one year and fail the next? How can a single plant bear the leaves of one species and the flowers of another? What makes some huckleberries red, others black? Is it true they grow better on slopes with a view? Are there really albino huckleberries, and what does it mean?

Many berriers have wondered these things for years, and the range of their answers is as broad as a mountain and deep as several valleys. Scientists have applied their special logic and supplied more answers. Still many questions remain. We can say the huckleberry is in some ways mysterious. In some ways unexplained. In some ways simply unasked. But secret?

If there is a secret about huckleberries, it is very badly kept. Who could keep so large a secret among so many people, especially when the truth is so welcome? If there is a secret about huckleberries, whole mountains are the clue and one mouthful is the answer. No, the real secret about huckleberries is that they aren't secret at all.

The only secrets of the huckleberry are the secrets of the huckleberry picker—and these may be many, and closely kept. Berriers live by few rules, but everywhere they observe one courtesy, sovereign and inviolable. That is: never ask where. Never, under any circumstances, look the slightest bit interested in where. If by chance someone should be drunk or careless or testing you and mentions a location by name, ignore it. Pretend you are at a formal auction well beyond your means. Act as if you didn't hear. Look away graciously, as you would from an open zipper.

And don't believe it. Don't believe one thing you hear about huckleberries, including this. The only thing wilder than a huckleberry is a huckleberry story; picking time is thinking time, and hours in the woods mean hours of stories.

There are the size stories: "Big as a golf ball! Big as both my thumbs!"; the adventure stories: "and then I saw the bear!"; the believe-it-or-don't stories: "white as snow, honest to god, but they canned up as sweet as any." There are the how-far-I-had-to-go stories and the how-lost-I-got stories, the best-pie-I-ever-had and the lousiest-picking-you-ever-saw stories.

As for these and all huckleberry stories, believe only what you see, or better, what you put in your mouth. When in doubt, remember that the human mouth is better suited to eating huckleberries than to talking about them.

And yet who can resist? Either huckleberries, or talking about them? Secrecy is one thing; silence, quite another. The topic of huckleberries is admirably suited to discussions of any length or level of seriousness, from a summer evening liar's contest to an earnest session of Sunday morning philosophy.

For the place of huckleberries in the lives of people who live among them is a place of honor, a place of great history. It is a place of grandmothers and first memories, of barefoot summers and sudden rain; it is a place of discovery, of being the first human to walk a certain path in a certain way under a certain sky, sure in the knowledge that no one has ever touched or tasted these berries before.

The place of the huckleberry is the wide world beyond the kitchen, the shouting freedom of whole days headed in no direction but your own. And it is the world inside the kitchen, too, the scalding and the scolding and the steaming and the smelling; the loving and the laughing and the tasting of all the higher fruits that simmer in the heart of the home. The place of the huckleberry is in the hands of women, strong and silent, tending the most basic needs in the most beautiful ways. The place of the huckleberry is in the hearts of men, joyous and grateful for all that is wild and all that is warm.

To live among huckleberries is a mark of distinction, the right and privilege of free people. To live among huckleberries is one last account held in escrow against the sweltering power claimed loudly by cities. To live among huckleberries is to touch a chord of timelessness, to know the wildness in the reedy backbone of our DNA. To live among huckleberries is—simply—to be alive.

How is living among huckleberries different from living among lowland blueberries or bramblerow raspberries? Since when is one wild fruit any more or less than another?

• Since flavor: since that intangible elixir, the direct wordless message from earth to mouth. Just as a tomato's taste tells of a life in hot sun and moist, black soil, the sweet sting of the huckleberry speaks of high mountain breezes and crisp, sudden nights; sweet, fast water, and fragile soil on strong slopes.

• Since price: since huckleberries defy the economic assumption that what is precious must also be rare. A classic American success story, the huckleberry has come from its remote and unlikely home to become one of the most highly valued wild crops on the market. Huckleberries are not packaged in bulk, labeled generically and distributed through IGA; huckleberries are cooked in small batches, labeled in gold and sold at exclusive art galleries in Florida. And yet where they grow, they grow within arm's reach of every trail. Where they grow, they grow by the acre.

• And since character. What other berry has remained independent so long and so well? What other berry has had the good sense to grow where and as the huckleberry does, in the fresh and scenic reaches of the continental northwest? Is there a berry which can match its grace and generosity, sustaining a threatened race of bears while clinging artfully to a mountain? Is there another berry as unassuming in demeanor, as understated in appearance? One as certain to call forth the attention of poets and the allegiance of generations?

No, the huckleberry alone can claim these distinctions, though it would never deny another berry its due. The huckleberry alone compounds its contradictions and shares its private world, revels in its own mystery and guards the secrets of others. Alone the huckleberry proves wildness to us, and proves it again, until we learn the language of our land, the flavor of freedom. Until we savor everything as if it were rare and perfect, everything a gift, everything precious and everything plentiful. Until we pick our days, too, with great joy, one by one, cherishing each and sharing them all. Until slowly we find ourselves at home with the huckleberry.

Huckleberries are precious and plentiful, wrapped in secrecy and everywhere in abundance. They come out every other year and sometimes in between, thriving in the forest where the trees let in plenty of sun. Picking huckleberries is good work for women, and no less for men. Huckleberries are blueberries and never will be; they are hard to look for and easy to find. Huckleberries are worth risking your life for, and one last, good reason for living.

Huckleberries are.

The Difference
Between a Huckleberry

Huckleberries and the Aurora Borealis

This year the crop was of such size and quantity all the bears in the world couldn't eat them up. My eyes bugged out at the size of the berries. Oldtimers have been telling me for years how big these huckleberries get: as big as a cherry, as big as a grape. Now I believe! In fact, I'll probably add a few tales myself: so big it took two bites to eat one berry, so thick the mountainside was black...

When I stood in awe like an ancient it was the evening of the day I had picked these big as a house huckleberries. Allison and I were camped on top of a mountain overlooking the reservoir. It was just a day or two past full moon so we stayed up to watch the moon shining on the water. That was enough beauty. We didn't need any more. But on top of that, suddenly the sky was full of light. It was the aurora borealis, the Northern Lights. Neither Allison or I had ever seen a really full display. Well, how many people have? We live in cities. But here we were, by pure accident on top of a mountain without a manufactured light anywhere, and on comes the aurora borealis. Shooting streamers across the sky. Waves of light undulating like the ocean. Clouds of light playing back and forth. We just stood there. We had to back up against a tree to steady ourselves so we wouldn't fall over backwards.

"What did the ancient Indians think of this?" I said to Allison. "What did the Eskimos think?" Wouldn't they think the world was coming to an end?

They didn't have encyclopedias to look it up and know it was electricity or sun spots or something. Actually, I don't want to know what the aurora borealis is. I just want to remember that on top of a mountain in August, 1982, I was in the middle of something awesome and it was a privilege to be part of it and I say Amen.

I stood in awe. Sometimes you have to do that. When you're in the middle of something so big you have no way of understanding it. Like an ancient primitive you stand there, just stand there.

—Kim Williams

If you want to call a certain berry huckleberry
and your neighbor wishes to call it a blueberry,
don't get excited. It is done everywhere.

—Kim Williams

What's in a Name?

Say huckleberry, and some will think of a great Mark Twain classic; others, a Saturday morning cartoon. Botanists might flip their mental notebooks to the section marked Heather, and anyone else might half-recall a bush they stumbled past, and berries they either did or didn't eat, in childhood days of tramping through forests and meadows anywhere.

A common name, as naturalist Kim Williams said, belongs to those who use it. In other words, a huckleberry is a huckleberry is a huckleberry.

But suppose you say huckleberry, and something happens. A face lights. Stories begin to flow. Then you know you have named one particular fruit on one particular plant in one particular part of the world, a berry that bears no relation to Huckleberry Finn, Huckleberry Hound, or the botanical genus *Gaylussacia*. And for those who have loved, lived with and sometimes for this berry, it is the only huckleberry that matters.

The Wild Mountain Huckleberry

Whortleberry, dewberry, bilberry, blueberry; mountain-this or thinleaved-that; big or blue or dwarf or globe, the huckleberry goes by many names. It grows wild, and only wild, in the remote areas of America's inland Northwest from Oregon to Alaska, the Pacific Ocean to the Continental Divide.

The huckleberry is a living exhibit of the colors purple can be, from deepest black to blues and reds. Smooth and round, about the size of garden peas, they are often described in terms of the hands used to pick them: "like the end of my little finger," or, "about the size of my fingernail." Any mention of thumbs indicates very good picking. Huckleberry bushes range from knee- to chest-high on a person, or haunch to head on a short bear.

In appearance the berry may seem somewhat unremarkable, and off the bush might even be mistaken for an ordinary blueberry. Once in the mouth, however, there can be no doubt: this is something else entirely. And here the trouble begins, because for those who have tasted mountain huckleberries, no description is needed; and for those who haven't, description is impossible. How do you describe the colors in yesterday's sunset, the smell of today's breeze, the taste of a waterfall?

One huckleberry cook knows better than to try. She says, "I just open a jar and hand them a spoon."

When Is a Huckleberry Not a Huckleberry?

Now the scientific believers out there are already waving a red flag. "That's not a true huckleberry," they chorus, "that's a kind of blueberry!"

A fair objection, since what we call a huckleberry is in fact a type of blueberry; and wildly unfair, since what we call a blueberry tastes nothing like a huckleberry. If this sounds confusing, good. It is.

Most botany textbooks will place the "true" huckleberry in the genus Gaylussacia. However, most of those texts will have been written on the East Coast, where *Vaccinium* plants are called blueberries and huckleberries belong to the genus Gaylussacia. These eastern huckleberries are also dark blue or purple, but they have larger seeds and less flavor than the mountain huckleberry, and are not usually considered a prime delicacy.

The western huckleberry is indeed of the genus Vaccinium and not Gaylussacia. Some purists remain firm that the name "huckleberry" belongs to the East, and the West will have to make do with "bilberries" or "dewberries." But most westerners have never heard of either bilberries or dewberries, and mountain people who have picked more berries than any five botanists will ever see have no problem calling a Vaccinium plant "huckleberry." So rather than defer to East Coast hegemony and unname the huckleberry, we shall stick with tradition and let the best berry win.

Huckleberries vs. Blueberries

As a member of the genus Vaccinium, the huckleberry is related to the wild blueberry of Michigan, Maine and other points east. In scientific terms it is equally close to the cultivated blueberry found in cans at the grocery store, muffins at the corner bakery and prefabricated snack pies all over the world. But here the kinship ends and botanical family ties come undone, as far as most berriers are concerned, because huckleberries and blueberries simply don't taste the same.

A true huckleberry advocate will usually dismiss blueberries, wild or cultivated, as "blahberries." Others say the two are as much alike as sweet cherries and pie cherries, or a fresh Granny Smith apple compared to last year's Golden Delicious. But blueberry fanciers are equally loyal, and will reject western varieties as sour impostors for the "true blue." So while a huckleberry may have more in common with a blueberry than it does with, say, a rocking chair, any educated tongue is going to know which you're putting on the toast.

Taste is not the only difference between blueberries and huckleberries. Most species of blueberries grow in bunches, and huckleberries singly, along the stem. Huckleberry bushes tend to have shallow roots with rhizomes, by which they reproduce. Blueberry roots go deeper.

However, there are species of huckleberry that bear fruit in bunches, and roots of the blueberry may grow deep or shallow depending on the soil and other local factors. Such blurred distinctions are quite common among the Vaccinium, causing endless annoyance for the scientist wanting to make firm identification of a plant.

The blueberry-huckleberry controversy may be best understood as a family quarrel. Both belong to *Ericaceæ*, the heath family. This designation tells a botanist that the plants have more in common with one another than they do with mint, geranium, cactus, palm, rose or other family groups.

The heath, or heathers, are woody plants with leathery leaves and bilaterally symmetrical flowers; all have twice as many stamens as petals, and upright anthers. Besides huckleberries and blueberries, the heaths include rhododendron, wintergreen, labrador tea, sand myrtle and trailing arbutus.

Within the large Ericaceæ family, huckleberries and blueberries are related most closely to one another, which may account for their sibling rivalry. Even Gaylussacia becomes a stepsister in their small sub-family. Should the logic of scientific classification have escaped you thus far, consider that the next of kin to both the huckleberry and blueberry is: the cranberry. Right. Botanically speaking, Thanksgiving-sauce bog cranberries are as much like the high mountain huckleberry as the store-pie blueberry.

So, for those who are still wondering, the blueberry is to the huckleberry as the cranberry is to the blueberry. Enough said.

The Cosmopolitan Cranberry

The name Vaccinium is derived from the Latin "vaccinus." This reference to cows is not widely understood, but cattle who come across Vaccinium plants will graze on them—and berries or no berries, the powerful flavor comes through the milk. As one rancher said, "It's the only way huckleberries aren't good."

Botanists call Vaccinium a "cosmopolitan" genus, meaning its species are found all over the world. The most common Vaccinium in Europe is V. myrtillus, the so-called bilberry, which is found at least as far afield as the Carpathian Mountains. English and Irish poets have sung and written about their fields of "heather," predominantly kaluna, and the Scots also have a Vaccinium they prize for its berries.

Varieties of cranberries, blueberries and huckleberries are found from Argentina to Siberia to Panama and back. The Vaccinium adapts well to a variety of growing circumstances, and has been known to grow to mythic proportions: in the southeastern United States, researchers have identified a single plant that is more than a thousand years old and covers a full square mile.

But few Vaccinium, while genetically equipped for such feats, opt for such size and age. Taken together with the wild and cultivated blueberries of the United States, one might conclude that the huckle-berry, as just another Vaccinium, is nothing special. Perhaps it is not.

But in discussing families and sub-families, genus and species, remember that these are based on morphology, which is just a fancy name for shape. In other words, plants that look like one another—in some very specific details of their leaves and flowers—are considered to be related.

But these botanical classifications ignore flavor altogether, and make no distinction between species that grow wild and those that allow themselves to be domesticated. There is no taxonomic considera-tion for a plant's character, history, or capacity for poetic inspiration. As far as pure objective science is concerned, the huckleberry might as well be just another cosmopolitan cranberry headed for somebody's table.

Lumpers & Splitters

So a huckleberry isn't officially a huckleberry, it isn't a blueberry either, and we know for sure it isn't a cranberry. Once you've gotten the courage to come right out and call a true Vaccinium a true huckleberry, you'd think the worst would be over, but no: the Vaccinium controversy rages on like an August fire through an inland forest.

The next task is to decide exactly which species each huckleberry represents. Since there are hundreds of species in the genus Vaccinium, the choices are many; unfortunately, among huckleberries the similari-ties are great and the differences few. Furthermore, here we come head to head with the character of the huckleberry itself: independence. Pure, feisty, unapologetic I'll-do-what-I-will-when-I-wish independence.

Taxonomists, the specialists whose job it is to put names on the huckleberry and other plants, must deal with this insubordination on a daily basis. A single huckleberry plant under scrutiny may exhibit three characteristics of one species, two of another, and then show a similarity to a third just to keep things interesting. It is rare enough for two specimens of any plant to be perfect, with no irregularities, and in Vaccinium it simply doesn't happen.

As earnestly as science attempts cold objectivity, the fact is that it is run by scientists who are all human beings and fully capable of personal bias. Taxonomists freely admit this, and in the spirit of their profession classify themselves as either "lumpers" or "splitters." A lumper looking at two similar plants may focus on their similarities and call them by the same name. Another equally qualified taxonomist, a splitter, would pick out subtle differences between the specimens and call them by separate species names, or at least different varieties within a species.

Whatever their individual biases, one thing on which most taxonomists agree is that huckleberries are difficult. There are two reasons for this.

One is that the huckleberry is highly climate-specific. In other words, huckleberries growing in the moist, low elevations of the Pacific

Coast are going to be different from those in the high, dry northern Rockies. Secondly—and this is also true of the wild blueberries of the East—the plants seem to cheerfully exchange genetic material among themselves, creating natural hybrids.

In the field it is next to impossible to tell which variations between plants are due to genetic differences and which are due to environmental adaptations. Since most people will find it easier to figure out which mountain range they are in—Cascade or Rocky—than to analyze huckleberry gene types, we will take the environmental approach and look at geographic differences in huckleberries.

Huckleberry Country

The wild huckleberry, which has resisted many attempts at cultivation, grows where it pleases. Usually it prefers forests and mountains some degree removed from civilization. Occasionally someone will report a patch of huckleberries "in the back yard," but usually these yards belong to houses that are extremely rural to begin with—backed up to public lands, for instance.

Depending on climate and growing conditions, one local huckleberry may be very different from another. No official survey has ever compared these variations across the northwest, but informal reports and isolated studies suggest certain trends.

The Pacific Coast, from the California line northward, is home to Vaccinium deliciosum, a dwarf plant with small sky-blue berries, and V. ovatum, the evergreen huckleberry used as decorative greenery in floral arrangements. Neither is much sought for its fruit—or as one picker sniffed, "Juice 'em!" The main Alaskan species is V. alaskaense. Other inhabitants of the huckleberry community, from British Columbia through the Cascades, are V. parvifolium, myrtillus, uliginosum, occidentale and ovalifolium.

But the most important species in the lower 48, from the perspective of the huckleberry enthusiast who picks to eat, are Vaccinium globulare and Vaccinium membranaceum. These two species are so similar that lumpers refer to them to as the globulare/membranaceum complex. However, each species seems to have defined its own turf: membranaceum favors the moist, milder climate of the Cascades and northern Idaho, and globulare prefers the dry, crisp atmosphere of the Northern Rockies.

Splitters will find some support for this distinction in the taxo-nomic history of the two species. V. membranaceum was first named by the Scottish botanist David Douglas, while collecting plants along the Columbia River through the Cascades. V. globulare was first collected from the Spanish Peaks area of Montana, in the Madison River drainage.

There are exceptions, of course. With the huckleberry there is always an exception, and globulare and membranaceum boundaries are no exception. Take the "superbush" found by a group of Forest Service botanists while doing huckleberry research in the Cascades. The winter had been unusually harsh, and most of the huckleberry fields had suf-fered dramatically. But there on a high ridge, smack in the middle of decimated membranaceum, was one bush in full glory, its branches loaded with literally quarts of plump, healthy huckleberries.

Curious about this extraordinary plant, the researchers took a sample to a taxonomist. The "superbush" keyed out as none other than Vaccinium globulare. Adapted as it was to the cold Rocky Mountains, to globulare what was a harsh winter for the Cascades probably seemed quite mild. The obvious next question, which has no equally obvious answer, is how that particular specimen of globulare migrated to its location. For now, this will have to remain one of the huckleberry's mysteries.

Apart from characteristics like frost-resistance, which cannot be seen by the casual observer, only fine distinctions separate these major species. Lumpers can resolve this as follows: membranaceum has pointed, "drip-tip" leaves, appropriate to the wetter climates; globulare has a more rounded leaf. Splitters will need to find a plant in bloom (no small feat, because huckleberry flowers come early and briefly), examine the shape of the blossom, and hope that the flowers don't key out differently from the leaves.

There is one other difference between globulare and membra-naceum, one that matters as much to taxonomy as taxonomy matters to your pie. That is flavor. Flavor differences correlate to the amount of water the plant gets in winter snowpack or summer precipitation. More water (membranaceum), more dilute flavor; less water (globulare), more concentrated flavor.

Which is best? Well, that's like asking which mountains are best: the Cascades, the Rockies, or somewhere in between. The easiest answer is that the best mountains are whichever mountains you're in, and the same is true of huckleberries. Any huckleberries are better than no huckleberries, so the best huckleberries are the ones in your mouth.

Other huckleberries grow along with globulare and membranaceum throughout the region. Two common species are Vaccinium caespitosum, the dwarf huckleberry, and V. scoparium, the grouse whortleberry. Caespitosum generally occupies the lowest elevations, globulare and membranaceum the middle montane regions, and V. scoparium grows well into subalpine terrain. Scoparium, the grouse whortleberry, is said to be the most common western Vaccinium, but as a huckleberry scoparium fails one very practical test: it is too tiny for sane adults to pick. Growing to the size of a BB at full maturity, it is usually left to the grouse for which it is named, and for whose dainty beak it is well suited.

Individuals with scientific aptitude, above-average curiosity, or spare time in the berrypatch, may amuse themselves sorting through these various species. But as huckleberry taxonomist Pete Stickney pointed out, the reason we have plant names is to communicate about them, and different people need different degrees of exactness in their communication. Someone trying to cultivate or hybridize a specific strain of huckleberry will need the most detailed genetic description available, right down to chromatograms and flavinoid analyses. The professional field ecologist can use a slightly less detailed description, and for 99 per cent of the rest of the us, one word will do:

Huckleberry.

Table of Huckleberries

Scientific name	Common names	Main features
V. globulare	blue huckleberry globe huckleberry	Flowers broader than long; rounded leaf tips; common to Montana
V. membranaceum	mountain huckleberry thinleaved huckleberry mountain bilberry big whortleberry black huckleberry tall bilberry thinleaved blueberry	Flowers longer than broad; "drip-tip" leaf ends; common to Idaho and Cascade areas
V. caespitosum	dwarf huckleberry	Shorter shrubs, under 18"; rounded stems (no ridge), obovate leaf shape
V. myrtilloides	velvet-leaved huckleberry	Distinctive hairiness to leaves and stems; common to Alberta
V. myrtillus	bilberry dwarf bilberry low bilberry	Green stem, small shrub, branching but not broomlike; lower elevations than scoparium
V. scoparium	grouse whortleberry	Tiny bright red berries, green stem, branching broomlike
V. occidentale	western huckleberry	no serration visible on leaf edges, even with hand lens; known to subalpine wet meadows near Idaho/Montana border

Adapted from "Field Key to the Vaccinium of Western Montana," by Pete Stickney, U.S. Forest Service, Intermountain Forest and Range Experiment Station, Missoula, Montana

Wild Mountain Heather

A Huckleberry Trip,
1930 Vintage

All of us three sisters were on our knees weeding the family vegetable garden on a hot July afternoon. While there working, the neighbor's boy walked by carrying a fishing pole and creel. He leaned over the fence and called, "I got a mess of trout up the creek and found ripe huckleberries along the bank!"

That message was all it took to get all three of us up from our weeding chores and run to the house to tell our mother the huckleberries were ripe. Then we started begging to go up the creek to where we could camp and pick berries.

We took orders from the neighbors, who said they would buy from us at the extravagant sum of 50¢ a gallon! The year was 1930, the economy was at a low ebb, jobs were almost impossible to find, so the idea of picking and selling berries seemed like a real bonanza.

The three of us could hardly wait until our father came home from work so that we could ask him if he would take us up into the mountains where we could set up camp for several days to pick berries. That evening before Dad even walked into the yard we met him at the gate, all asking and begging if he could take us, our picking buckets, and camping gear up to the trail. We all talked at once but somehow he got the message that the huckleberries were ripe and we wanted to pick them.

With a big smile he agreed he would take us up to the hills in the morning. Morning meant 5:00 A.M., as he had to return to town in time to get to work at 8:00. We were so excited that we hardly had time to eat dinner as we planned what to take: food, canteens, bed rolls, canvas tarps, picking pails, pots and pans, and pack boards with old five gallon oil cans strapped to them. We sorted out our hiking shoes or old tennis shoes, heavy socks, jackets, straw hats, denim overalls and shirts to wear while berry picking. Along with these items we packed an axe, shovel, soap, towels, matches and tissues to make life more bearable at camp.

All this we stacked in Dad's old Model T truck that evening so that we would be ready to roll at the crack of dawn.

Early the next morning we all climbed into the back of the truck and headed for the hills. And they were hills! One required all three of us girls to get out and push the old Model T most of the way up as it wheezed and almost died in its effort. After our pushing and two or three heroic efforts on the car's part, we made it to the top. Luckily the rest of the trip was gentler and we got to the site by the stream where we would camp.

Dad helped us unload the boxes of food and our buckets and bedrolls. We didn't have sleeping bags, so our bed was made up of cedar boughs piled on the ground. An old army blanket of WWI vintage was put over the boughs with another blanket on top. Overhead we hung a canvas tarp to protect us from falling tree cones or possible rain. Then Dad said goodbye to us and wished us luck with our berry picking. He promised to return the following evening to take home the berries we would pick; Mother had promised to deliver them to the neighbors.

As we watched Dad and the old truck raise dust along the road in their departure, we hurriedly put away our provisions. The milk in glass quart jars we anchored in the creek to keep cool and fresh. The pounds of butter we put into small cloth sugar sacks with rocks in the bottom and placed them alongside the milk. The rest of the food we stored under wooden apple boxes weighted down with heavy rocks. Such coverings we hoped would discourage varmints and other critters from eating our food stores.

We cleaned out an area in front of our bed site, scraped duff and dirt away and lined up rocks in a circle to be our fire hearth. With these preliminary chores done, we hurriedly put on our backpacks and gathered our picking buckets and water containers. Mother had made us a lunch of sandwiches and fruit and peeled vegetables. We tossed the lunch in the can on the pack and in single file started walking up the forestry trail.

The trail was narrow, covered with leaves and evergreen needles. As we went further up the trail, more craggy rocks appeared with more bare areas of dust. We watched for sightings of huckleberries. Up, up and away we hiked, stopping for a moment to pick a few berries from the scattering of bushes along our way. Finally we rounded a bend and up above the trail spread a clearing of about an acre. There were huckleberry bushes with ripe berries hanging from every bush.

We set down our packboards and large cans under a tree. Then we hung a big red bandana handkerchief on a branch to mark the place where our packs were located.

We picked berries into buckets hung from our waist by a belt; this way both hands could be used for picking. Each time we filled the small bucket we would return and put the berries into the large cans on the back pack. We set a goal to pick five gallons of berries apiece before returning to camp.

We busied ourselves picking and tried to stay fairly close together. During the afternoon I could hear my sister picking near me and I asked how the berries were where she was. She did not answer me, but I could still hear her shuffling around in the bushes so I said, "Well, if you won't tell me I'm coming over to find out for myself."

To get to her, I climbed over a big windfall tree and in doing so rolled over and fell on my back. When I looked up from the ground, it was not my sister I saw but a big black bear who had been the one sharing the berry patch with me. I let out a horrendous yell and the bear turned and headed downhill at full speed. In his dash to get away, I could see the soles of his back feet that looked like moccasins.

My sisters were close by, so we assured each other that no harm would befall us and continued to pick berries until all of the five-gallon cans on our backpacks were filled. Then we put the packs on our backs, took a long swig of water from our canteens and finished eating a sandwich as we hiked back to our camp.

The walk back to camp seemed much longer than the three or four miles up. The five gallons of huckleberries became heavier with each mile.

Back at camp, we cooked hamburgers over the campfire and later we bathed in the cold stream. Finally we climbed into bed to dream of our newfound wealth that we would earn by picking and selling huckleberries that summer.

—Angela Holiday

...as we had nothing but a little flour and parched meal to eat except the berries with which the Indians furnished us I directed Drewyer and Shields to hunt a few hours and try to kill something, the Indians furnished them with horses...about 1 a.m. the hunters returned had not killed a single Antelope...I now directed McNeal to make me a little paist with the flour and added some berries to it which I found very pallatable...

—Captain Meriwether Lewis while waiting for Clark after crossing the Continental Divide, August 14, 1805

A History of Berrypicking in America

Long before Meriwether Lewis had his cook improvise some berry pancakes to stave off starvation a few more days, the wild blueberry held an important place in the economy of the North American continent.

The most complete history of huckleberries in America was prepared by Henry David Thoreau. He found references to huckleberries as early as 1615 in the journals of Champlain, and later in the records of other 17th-century explorers. It was Thoreau who identified the English author John Lawson as the first to use the word "huckleberry" while writing about North Carolina in 1709.

Throughout New England, English settlers bought blueberries from the great quantities harvested and dried by the Native Americans. After Lewis and Clark's observations in the West, 19th-century explorers in Wisconsin and the upper Mississippi River drainages also noted Native American traditions of picking and drying wild blueberries. During his Civil War era search for a railroad pass over the Cascades, George Mc-Clellan also encountered tribes who relied heavily on huckleberries.

Native American Traditions

As Thoreau said of the Native American tribes regarding huckleberries, "I will go far enough back for my authorities to show that they did not learn the use of these berries from us whites."

In the Cascades, huckleberries have been an important element in Yakima life and culture. Picking into handmade wickerware baskets, the Yakima preserved huckleberries by dehydration, drying the berries near a burning log rather than attempting to sun-dry them in the humid Cascade summers.

The huckleberry territory of the northern Rockies is the traditional home of the Kootenai and Salish tribes. The wealth of the land sustained the people with fish, game and roots as well as different kinds of berries. As an abundant and highly nutritious resource, huckleberries have held an important place in Native American culture, along with serviceberries and chokecherries.

In the days of her childhood, Salish elder Agnes Vanderburg remembers, her people began the huckleberry season with an important ceremony. During the tribe's annual July powwow, the chief sent two sisters up into the hills to pick huckleberries while everyone else waited in camp. When the sisters came back with their two big buckets full of berries, everyone gathered.

Prayers were said, and then the buckets were passed through the crowd. Each person took two berries and waited while the bucket continued its rounds. If little Agnes looked like she was getting impatient to taste her first berries of the season, she got a sharp elbow in the ribs from Mom; only when everyone had two huckleberries was it time to eat.

No berries could be eaten before this time, lest a person choke on them. After the ceremony, and only then, was it safe to pick berries of all kinds.

According to early anthropological reports, the Salish "First Roots" ceremony was similar. Two women would go to a known field of bitterroot, a tribal food staple. The older woman would pray for "success, security, good health and fortune for all," first invoking the sun and then the earth with the same prayer. The women then gathered roots and brought them back to be prepared for the community. It was understood that unless this ceremony was observed each year, the crop would be "wizened and scarce."

Kootenai culture also had a celebration of huckleberry season, the Grizzly Bear Dance. It was second in importance to the annual Sun Dance, which expressed the tribal knowledge that "nothing would bring forth upon the earth without the Sun's efforts." Anthropologists say that no meat—not deer nor elk nor bear itself—had the ritual significance of huckleberries in the Grizzly Bear Dance.

The dance was held at the start of the berry season, beginning with the building of a large dance lodge especially for the occasion. Inside, religious leaders created an altar using the skull and front paws of the Grizzly Bear, while the community's adults waited quietly in their tipis. When the lodge was ready, assistants collected the medicine bags of each member and brought them to the altar; when all bags had been collected and arranged, the adults would come to the lodge with their smoking pipes.

The ceremony incorporated dancing, singing and the telling of medicine experiences. Pipes were lit by the head shaman from a central fire, with a prayer for each person "that a plentiful supply of berries will be found both during the current year and the next, and that Grizzly Bear will help him in his troubles." The dance continued for three days, ending at dawn with a feast on the berries they believed the grizzly bear "most enjoys."

Native American Uses

As a rule, men were responsible for gathering animal food by hunting and fishing; women collected roots and berries and other vegetable staples.

Berry gathering was done by family groups, with each family returning to its area every summer. Sometimes the trips were combined with fall hunting: men went for the fattened deer and elk, and women for the ripe huckleberries and serviceberries. The berries were gathered in bark baskets and consolidated in hide parfleches to be carried by horses to the lodge. Any berries not eaten fresh could be spread on hides to dry in the sun, then ground or stored whole. The fruit might also be mashed into cakes and dried in the sun.

Although the local people knew of the Plains Indian practice of mixing fruit with meat for a winter staple known as "pemmican," consensus is that neither the Salish nor Kootenai adulterated their berries in this way.

The preserved huckleberries and serviceberries were an important element in the winter diet. At least one feast dish used dried huckleberries: a stew of cooked bitterroot and berries, possibly with venison if it was available.

Some say huckleberry tea was made for medicinal purposes. It was brewed from the roots and stems to treat ear or kidney disease, and from the leaves for arthritis and rheumatism.

These Native Americans did not draw sharp lines between economic, social and religious activities, and berrypicking, like everything else, had multiple significance. In a single excursion a woman might provide winter food for her family, visit with a cousin from a nearby camp, and have a medicine experience. One Salish man remembered it was in a berrypatch that his guardian spirit revealed itself to him when he was a young boy:

"When I was about nine years old I went huckleberrying in the Mission [Mountains] with my mother and sister. They left me all alone one day; night came on; I cried and cried, I was so much afraid; afraid Bear would bite me, afraid Coyote would bite me. I cried until maybe four o'clock in the morning, when I heard the singing of this spirit. I cannot explain the name. You know in a burn how the dead trees rub together and talk? This is the spirit of the Wind-in-Dead-Timber speaking. He told me I was soon to be a man and would no longer be afraid; I would always be protected if I sang this song which he taught me. I used this song all my life. Sometimes I used it for curing other people who had been injured."

Early White Settlers: Edna's Story

The earliest arrivals from the East found their lives shaped by the land much as the Native American people did. Supply lines from the cities were few and fragile, and the money to exchange for goods was even rarer.

Edna McCann came to western Montana by covered wagon in 1905. The family lived in a log home on a homestead of 160 acres. At first their only lights were mining candles; after a year they finally got a kerosene lamp. A team of horses was used to haul water from a nearby creek in barrels on a wagon or sled, depending on the season. Everything the family needed they grew or harvested themselves, or traded for; Edna isn't even sure how they got the money to buy clothes in the early days. When she was 10 her father bought a sawmill and cut ties for the Great Northern Railroad. "Then we were rich!" she remembers.

As one of the few native fruits, huckleberries were a serious part of the McCann family diet. Edna's father was in charge of picking, and her mother, preserving. As soon as the children were old enough to climb around the mountain, they would ride with their father into the forest, tether the horses, and pick huckleberries. The Forest Service had not yet made its mark on the land, and the woods were thick timber, big timber, at least until the 1910 fire. The only trails were the ones made by deer and bear; even elk had not yet been introduced. Still, "you couldn't lose me in the mountains!" Edna remembers proudly.

Indians from a nearby camp also came to pick huckleberries. They gathered the berries in what Edna thought looked like blankets, and carried them back home tied to their horses.

Mr. McCann had a rule about berrying. Edna and her brother each had a quota, to fill a five-gallon lard can before they came down off the mountain. Edna picked hers into a small metal pail that she used to fill the large bucket.

When all the lard cans were full, they would go back to the horses and ride home to where Mother was waiting with the canning supplies. On their monthly trip with the team to the local store, she had bought sugar and glass Mason jars in preparation for berry season. Since they had picked the berries by hand there were few twigs or leaves to remove, and the fruit went straight into jars.

The canned berries were stored in the root cellar that Edna's father had dug by hand, where there were shelves for jars and bins for potatoes and carrots. Each year the family tried to anticipate how much fruit they would need to get through the year, and continued canning until the shelves were full of huckleberries and raspberries.

All of this earnest domesticity was going on at exactly the same time that the Wild West was regaling the East with tales of shoot-'em-ups and gunfights. In Edna's town were three gunmen "who might have had bad names, but still respected a woman." These were fellows who had once ambushed a certain "house" in town, a house at which some of their friends had contracted an unpleasant disease. To redress their grievances they rode around and around the house, shooting out lights and into the air until the proprietress came out with her bags packed and headed for the train depot. And yet these were the same men who took a protective watch over little Edna, which meant that nobody said anything cross to Edna McCann—at least not in public, and certainly not when they were around.

The gunmen are gone now, and so, in Edna's opinion, are the "really big patches" of huckleberries. Whether the grounds have been overpicked by commercial harvesters, overmanaged by the Forest Service, or simply overgrown by the forest canopy in the decades since the big burn of 1910, as far as Edna is concerned big berrying days are a thing of the past.

The Old Days

Yes, it was different in the old days. The modern conveniences of automobiles and roads—not to mention cooking and canning equipment—changed huckleberrying a great deal. It used to be that anyone who didn't live right on top of a huckleberry patch had to travel long distances on treacherous trails to get to the berries. One woman remembered, "Sometimes that meant going in the buggy, or later years the Model T. When Papa went along it was always a picnic…"

Lunch was usually simple, and huckleberries were always the main course. A family that kept cows might favor the berries with a pint of fresh cream, some sugar and a bowl; others would pack along two slices of bread, buttered and sugared and ready for fresh berry filling. Taking horses had some advantages. If it took a long time to fill the five-gallon cans, it was nice to lift tired limbs onto a horse that knew its way home in the dark.

The best picking was usually in roadless areas, which meant that in order to get any huckleberries, "you had to be able to climb a mountain." It was often the younger generation's responsibility to bring back extra berries, enough for the older generation and perhaps a few more to sell on the side.

Huckleberries have been a good cash crop as long as anyone can remember. Even when the prices were 35 or 50 cents a gallon, a child could pick enough berries to buy school clothes or shoes. In later years when prices went up, a person could climb a mountain, pick two gallons, come back down and sell them for a dollar apiece and feel that the day had been very productive.

Most huckleberrying gear was designed and invented with available materials. The five-gallon lard or honey bucket was a standard feature of early huckleberry picking. With reflector ovens, store-bought or home-made from scrap metal, pies could be baked in the woods using the heat of the sun. One woman remembers her mother making a special picking apron for her father. It was made of heavy material with a wide hem and rounded edges. The top was tied at the waist, and a young willow branch was threaded all the way through the hem. The ends extended out behind the man's back and were tied to make a full circle. He held the skirt out under the bush, using a canvas paddle like a large tennis racket to beat berries into his apron.

Huckleberry Camps

Without other sources of fruit, huckleberrying was a critical economic activity. A family could use up to a hundred quarts of huckleberries before the long winter was over. Finding, picking and canning a hundred quarts of berries was a lot of work even in a good year. Instead of making repeated trips, a family who lived in town or far from good berry patches could simply pack up, head into the mountains and set up camp. For their children, memories of summer would be inextricably twined with memories of "huckleberry camp."

Huckleberry camp might last a few days, a week, six weeks or two months, depending on the crop and the inclinations of the family. It was a tradition on a par with the fall hunting camp, except the children didn't have to be taken out of school for berrying. Every family had its own style and traditions, its own preferred approach to the huckleberry patch, but the basic elements were the same.

At some appointed time in late July or August, the family packed up a truck or team with berry picking and canning and camping gear, and headed for the hills.

Central to the operation was the canning equipment. This might be a large kettle to boil over an open campfire, or even a special stove that could be packed in on horseback, along with the jars and sugar needed for preserving. The design and choice of equipment was always distinctive from family to family, based on individual preferences and the materials at hand. In the 1920s, Forest Service workers had access to unlimited coal-oil cans, a rectangular shape that could be used as a picking container or made into a reflector oven. The same workers used canvas and telephone wire to make beaters and pickers with which they could net as many as 15 gallons a day.

Children would be enlisted for the picking, which they liberally mixed with playing, especially on outings with other families that had children. Fathers might lead the picking expeditions, unless they stayed behind to work in town or tend to the farm. Mothers were usually in charge of the canning operation.

Like picking gear, cleaning apparatus was homemade from available materials. Families who picked by hand had little cleaning to do, but those who used pickers or beaters would collect twigs and leaves along with the berries which had to be removed before canning. Most cleaning rigs used some kind of a long, sloping trough of canvas, piano wire, or screen—any material that would allow the berries to roll while catching or dropping the twigs and leaves. At the end of the chute might be a wool blanket stretched tight, off which berries would take one final bounce into a box, leaving behind the last bits of debris. Children who weren't picking or playing might be assigned to sweep the chute clean with evergreen boughs.

There were plenty of adventures to be had in huckleberry camp, far from the comforts of civilization. For one thing, huckleberry country was always bear country. But stories of bear conflicts are few in the old days, when there were relatively few people infringing on the bear's territory. Back then bears were considered natural companions in the berrypatch. One man remembered seeing a bear right in the picking grounds he'd been heading for. Instead of turning back, Tex said, "I figured he knows where the berries are, I'll just go up and help him."

That strategy wasn't always workable then, any more than it is now. One woman patiently worked around a bear whose patch she was sharing, but soon the bear found it was easier to eat out of her full buckets rather than pick his own off the bushes. The woman finally moved down the road several miles—and two days later so did her bear. At that point she decided it was time to break camp and go home.

The Legendary Camps of the 1930s and 1940s

There is a certain drainage in western Montana that is notorious for its huckleberries. It has the reputation for producing the most berries, the most consistently, of any area around, ever. Tall claims. But who's to argue?

As huckleberries and stories proliferate hand in hand, the stories about this particularly famous area are many. It was such good picking that during the 1930s and 1940s, much of western Montana converged on this area and set up huckleberry camp. On one side of the road the Native Americans would have as many as 500 tipi lodges; on the other side of the road would be a similar encampment of whites.

A tarpaper store was improvised at the camp, from which the pickers could buy basic necessities. Pop and milk at the store were kept cool by placing them in the creek. One woman set up a bread-baking operation and sold fresh bread to the people in camp.

Those years produced boxcar-loads of berries, they say, and at least as many stories. At times there would be so many pickers in the woods that buckets could be seen glinting in the sun across the canyons. At times there would be so many berries that it was hard for horses to pack them all out.

The big huckleberry camp had a boom-town atmosphere, rough and ready for whatever came next. There were stories of drinking and dancing, Native Americans gambling on the "stick game" late into the night, and plenty of vehicles off the road by morning.

Perhaps it was Prohibition that spawned the rumor about a mean old cuss who brought in anti-freeze and sold it as liquor. Some people died from that one. Word had it his whole clan was just as mean and the old man got his share at least once by getting beat up during a family reunion. The story goes that his kids once robbed the camp store—but never could spend the money because they knew what would happen when they got found out. They hid the loot instead, and to this day somewhere out in a huckleberry patch is a buried treasure of 21 silver dollars.

Huckleberry Jack

Not all picking was done for exclusively personal use. Especially in a good year, a family could collect far more huckleberries than they and their relatives could consume. Any extra they sold commercially. In one Montana county, the local newspaper estimated that the huckleberry industry would gross a whopping $50,000 in 1936. Added to the mix of pickers, then, were the brokers, who would buy huckleberries at the big camps and ship them out by truck or train to markets in the city.

One such broker was Jack Laderer. Francis Pettit of the Forest Service remembers "Huckleberry Jack" as a wiry, rugged old bachelor who worked part time for the government. In 1922 Jack took his first assignment as a summer fire lookout—a station that suited him fine, because he didn't care all that much to be around a lot of people any-way. Laderer worked 20 summers on lookouts in western Montana before turning his efforts toward huckleberries.

He got the name "Huckleberry Jack" while on the lookout. He liked to add huckleberries to his muffins and hotcakes, and his reputa-tion as a cook was such that people would walk all the way up to the lookout just to eat his hotcakes.

For 30 years Huckleberry Jack was a fixture around the great camps of western Montana. Once a week he bought a load of berries, hauling them into town in a surplus WWII truck. After the early 1950s he bought a Volkswagen to replace the truck "because it was too slow for him."

Jack ran his operation from a base on the outskirts of one of the big camps. In later years this violated the Forest Service regulations that prohibited camping more than two weeks in one spot, but Jack had his connections. He developed a semi-official status for himself by maintaining weather records for the Forest Service, helping to plan fall burning, and more or less keeping an eye out for trouble at the big camps nearby. He put the box of weather equipment in front of his camp with the large "U.S. Property" sign facing outward, giving his operation the look of authority. And there he stayed, two weeks or no. Even the administrators had to admit he had been here before they were.

Each day Jack would hike five miles into good berry ground, pick five gallons of berries, and be back at camp around two or three o'clock. After a nip of wine he would build a fire and set up a kettle on a tripod to heat. After lunch, usually vegetable stew, Jack would clean and then cold-pack his berries over the open fire.

After the camps broke up in the fall and there was no more brokering to be done, Jack finished out his season by taking canned huckleberries to Oregon to sell. Later in the winter he would return with a load of fresh fruit to sell in Montana.

Changing Times

Over the years, the old timers say, things have changed a great deal. The roads, the camps, the picking and canning equipment; the cars, the traditions, the families. Even the picking grounds have changed as timber has grown up over old burn areas that used to be the best berry patches. But one thing hasn't changed, they say, and they don't expect it to:

"The huckleberries."

The Secret Life
Of the Huckleberry

Say Huckleberry, But Keep
Your Lucky Star Secret

Plink: like meteors colliding light
years away, the first berry—
steel ball bearing, big
as the tip of your little finger—
rings its purple reflection
around mirror walls
of a 3-pound Hills Brothers tin—
every bird and squirrel alert. One
lone berry, single planet
in deepest space, makes you swear
you'll never see home again.

But ah, the magic, the magic
of double-aught buck, loaded
huckleberry bushes hanging
like galactic mobiles
steep into your secret patch. So easy
to find yourself lost
at gravity zero in fantasy
or future, all fingers and eyes
on automatic pilot, the heart
tagging along on cruise. Go ahead.

But don't spill the berries.
Your only claim to fame and fortune,
this instinct to fill space. Daydream
new haunts, recaptured friends
and family gone years ago. Picture
close calls with frontier black, comet,
Great Bear. Nothing snaps you
out of picking this patch clean.
Pupil to pupil, hand over sprig,
these bushes lure and lose
you into orbits, up, over, and around
the mountain, the universe—
this brim of your bucket.

—Paul Zarzyski

Why the wildflower? I don't know. But I know the wild-flower, and I know why.

A Reluctant Subject

Secrets again. It's not that the huckleberry has gone to any great length to keep things from people; it's just that people, for their own reasons, haven't gone to the same lengths in studying huckleberries that they have in studying, say, timber.

Admittedly, the huckleberry is a difficult subject. Its chosen habitat rarely extends to scientific backyards and other easily-observed experimental plots. Unlike more submissive species, it does not withstand transplanting well, and is less than enthusiastic about being propagated from seeds or cuttings. Still the huckleberry receives considerable attention, informally on the part of individuals and in the more systematic ways of scientists and entrepreneurs. Many people who have planted a berry or a cutting or a bush, only to watch it fail, doubt that the species will ever be domesticated, but the value and charm of the huckleberry are such that efforts will surely persist until this is accomplished.

While little is yet known in purely scientific terms about life and times of Vaccinium globulare, the accumulated wisdom of generations of huckleberry pickers and eaters is considerable. Combining this with the available scientific inquiry, we can gather some insight into this shy and effusive, wild and reclusive plant.

Huckleberry Production

To the delight of poets and the dismay of nearly everyone else, huckleberries are highly unpredictable in almost every regard. In a century devoted to highly predictable systems and error-free computers there is something refreshing about a perpetually unknown quantity. In both where and how they grow, huckleberries have earned this mystique.

Veteran berriers may have a favorite patch or two that yield consistently good quantities, but most will agree that the crop is different every year—often dramatically different. Where a north drainage was great last year, the south may be exceptional this year, and vice versa; then a year may come when nothing looks very good until you get up above the 5,000′ level, at which point all the fields explode into the biggest berries anyone has ever seen.

It may be that the huckleberry is just fickle by nature, which would be in character for a plant that resists every other kind of pigeonholing. On the other hand, the huckleberry may be very much a creature of habit that only seems unpredictable to human beings who haven't figured out exactly what those habits are.

Huckleberry Habitat

Like everyone else, huckleberries cannot be understood apart from the place they come from. The berry's context is appropriate to its character: wild, remote, slightly dangerous, exquisite, irreproducible. In these respects, the high ranges of the Northern Rockies are ideal huckleberry habitat.

Like other plants, Vaccinium globulare has requirements for growth and production. It has certain preferences for elevation, moisture, soil, slope, aspect to the sun and microclimatic factors. However, unlike indicator species such as queen's cup or beargrass, which are found only in one particular forest type, the huckleberry is widely adapted across several ecosystems. One exception is that the berry is generally not found in dense stands of spruce/cedar habitat because the leafy cover doesn't allow enough sun to the forest floor. Otherwise the huckleberry is a genial member of the forest community and coexists happily with all manner of other species, as long as its particular needs are met by the habitat.

Truly, the amount of terrain that meets those needs is vast; huckleberries are found in profusion throughout the mountains of the inland Northwest. In general, prime huckleberry habitat will have the following characteristics:

• Elevation: As any picker can prove by rubbery legs at the end of the day, the huckleberry prefers relatively high elevations. Depending on other factors, those elevations run from 3,500′ to 7,000′. The crisp alpine nights seem integral to the berry's cycle.

• Moisture: Huckleberries require a consistent supply of water, from rain or soil supply or underground sources. They are rarely found in dry or grassy locations.

• Aspect: While moisture is essential, huckleberries do not like wet feet. For this and perhaps because they like the view, they tend to gather on slopes rather than flat places. A hillside of 25° to 30° is optimal. Opinions differ as to the best aspect of these slopes—north, south, east or west—but generally north and east slopes will be productive if they get sufficient light, and south and west slopes if they get sufficient moisture.

• Soil: The acidic soils typical to coniferous forests are most suited to the huckleberry. It requires moist, deep soils that never dry out, and the best growing conditions have loose, lightly aerated "ash cap" soils, the kind generated by volcanic eruptions. Many plants will be seen growing up and around fallen timber; they appear to take nutrients from the rotting wood.

• Light: Huckleberry bushes can and do grow under a forest canopy, but they fare much better in partial shade or on well watered or sheltered open slopes.

• Weather: Despite its wide adaptability to habitat, the huckleberry remains highly sensitive to its surroundings. An otherwise prime site might yield nothing if it falls in a "frost pocket" with poor air drainage resulting in excessively cold nights at sensitive times. On the other hand, an unlikely flat spot along a shore may host a crop of productive bushes because of the lake's moisture and temperature buffering. Microclimatic changes—from year to year, as well as location to location—have enormous impacts on the berry crop, if not on the survival of the plant.

• Companion vegetation: As noted, the huckleberry associates quite freely with other plants. However, because of its need for light, the best huckleberry sites are often moderate to poor timber sites. Grand fir and alpine fir are frequent companions to the huckleberry, as are serviceberry and other fruiting plants.

The Fire Factor

In a typical year, huckleberries and wildfire both peak during the dog days of August. But the relationship between fire and Vaccinium goes much deeper than the coincidence of summer, as berriers and scientists both acknowledge. Exactly what that relationship is remains an object of study and speculation, providing plenty of grist for the opinion mills at the kitchen table and on the back porch. For the unconvinced, a quick comparison of historic burn areas with traditional berrying grounds in Idaho and Montana should be enough to prompt a little headscratching.

Fire has been known to enhance—and some believe it is necessary for—optimal huckleberry habitat. The same acidic soils the huckleberry favors also support the coniferous forests that are prone to fires. On the other side of the ecological equation, the huckleberry's extensive rhizome systems and its relatively rapid regrowth after fires give the plant an important role in controlling erosion and redistributing nutrients through a burned area.

Eastern blueberry fields are deliberately burned every other year to stimulate production, but here again the huckleberry differs from its cousin. Perhaps because of drier conditions in the west, the huckleberry does not respond nearly as favorably, directly or quickly to fire. One study suggests that spring burning—when the moist ground insulates root systems from heat damage—is beneficial to the huckleberry, while fall fires damage the rhizomes and have a negative effect on the plant itself.

Whether or not burning directly affects the berry bush, a forest fire accomplishes one end critical to huckleberry habitat. In dense woods, the fire cuts away some or all of the overstory, which can deprive the berry bushes of necessary sunlight. Also, the burn releases an enormous amount of nutrients into the soil from the trees and other brush, creating the nutritional balance on which huckleberries thrive.

The relationship between fire and huckleberries is best seen not under a magnifying glass in experimental plots followed over one or two seasons, but in a satellite view of many acres over decades. After a burn, it takes many, many years for the forest to restore its canopy. It is during this time that the huckleberries are serving their ecological function of erosion control and nutrient transport, while also producing berries— like crazy.

Rumors to the contrary, there is little evidence that native people practiced intentional burns on huckleberry fields. In any case it would not be an effective short-term strategy because of the relatively slow growth rate in northwestern forests. After a major fire, it can take 15 to

20 years for optimum berry production to return. Even in the Oregon Cascades, which typically promote more rapid growth than the Rockies, scientific studies have shown that V. membranaceum did not bloom for two years after a fire, and berry production still had not been fully restored more than seven years after the burn. Once reestablished, though, the huckleberry fields flourish for the next 75 or more years, or until the tree canopy closes overhead.

The primal relationship between fire and huckleberries raises interesting questions for forest management. Fire suppression, widely practiced to preserve timber, may pose a significant threat to huckleberries. The real danger is that the incremental losses may not be noticed in the short term. If decades of fire control have not noticeably hurt the huckleberry crop in Montana—and this has not been measured—any sense of security must be balanced by the awareness that much that crop is still coming from a vast area devastated by the cataclysmic fire of 1910, the area known as "the 1910 burn." If that region continues to grow over without new terrain being opened up, the effects of fire suppression will soon be felt in lighter buckets and longer faces on berrypickers.

It may be that future huckleberry production will depend on thoughtful woodcutting practices in place of natural fire. Berriers know that good fields can sometimes develop in logged areas as well as in old burns and avalanche chutes, but the few scientific studies on logging and huckleberries show risks as well as benefits. Berries do respond favorably to the removal of the large tree overstory, but they cannot tolerate much soil disturbance. On clearcuts and scarified areas, the berry plants may take 50 years rather than 15 to regain their foothold.

"When It Was Good, It Was Very, Very Good..."

Within the large-scale variables of acceptable terrain and historic fires, the crop in an particular berry patch will vary from year to year based on factors that are still not well understood. Weather is the most obvious influence, but there also may be a secret relationship to sunspot cycles or volcanic activity in Hawaii, for all anyone knows.

Traditional lore says that berrypicking is best in alternate years: a good year followed by a bad year, and bad followed by good. Nellie Stark, a scientist who has studied huckleberries, believes that the rhythm of productivity is related to elevation, along with soils and climate. She estimates that true "bumper crops" come to lower elevations once every 10 years, to middle elevations every five or six years. Above 6,000' feet, she believes, nearly every year can be excellent.

"...And When It Was Bad, It Was Still Pretty Good"

The first weeks of July, rumors fly thick and fast through huckleberry country: this is a good year, a bad year, or a great year. Because of the erratic pattern of berry production, reports rarely agree. Someone whose favorite patch did well will report a bumper year; someone whose hunting grounds were frosted out will turn thumbs-down on the whole season.

Before making pronouncements about the huckleberry crop of any given year, one should carefully consider the enormity of the topic. First of all, no one even knows how many acres of the Northwest actually grow huckleberries. Forest Service biologist Don Minore, who spent years studying the berry fields of the Cascades, estimates some 100,000 productive acres in Washington and Oregon. As for the vast ranges of Montana, Idaho or Alaska, no one has even hazarded a guess.

As with every other story out of the huckleberry patch, annual reports on productivity must be taken with a grain of salt. The first returns, the ones that set the tone for the year, usually will be based on the most accessible picking grounds: the lower elevations, which are most inconsistent from year to year. But not even the oldest old-timer can remember a year when the huckleberry crop failed entirely. Experienced berriers know this, shrugging off most second-hand reports and early declarations in particular. They will wait and see for themselves.

The Sex Life of the Huckleberry

With all of these dangerously delicious fruits going into hands and mouths, not to mention beaks and claws, how do huckleberries ever reproduce? After all, the ostensible purpose of seed-bearing fruit is to drop to the ground and perpetuate the species.

But it turns out that once again the huckleberry chooses an independent path. As far as anyone knows, it rarely reproduces sexually (by seed). Instead the plant depends on its intricate rhizome system to send up new stems. Simple observation in the huckleberry patch will find very few actual seedlings sprouting underfoot; most of the stems come out of a complex network that can follow a tortuous path under fallen trees and other vegetation, making it difficult even to trace the primary root.

Nellie Stark thinks that, perhaps every hundred years or so, conditions will be just right for seeds, and major propagation will take place. Even then, those tiny seedlings must survive three to five critical years before establishing themselves as bearing plants.

A Year in the Life

How can the huckleberry can be so erratic from patch to patch, and yet so dependable from year to year? Some clues may come from a look at the sensitive points in an individual plant's life cycle. While the true secrets of productivity have not yet been revealed to mere mortals, the huckleberry still is a green plant with basic annual requirements for growth and fruiting. When these needs are met, one may expect a good crop; when any of these factors is absent or impaired, one may expect a corresponding loss.

Let's begin in the fall, after the berries have been picked or fallen from the bushes. In September or October, depending on the year and elevation, the huckleberry bushes will slowly turn a lusty red. This is the anthocyanin dye, which protects the chlorophyll, showing through as the plant's metabolism slows down. Soon the leaves fall.

By this time, the snow should begin to settle in the higher elevations. Contrary to its popular reputation as a cold and a hostile element, snow in the high country is actually a critical protector. Like the big "blanket" of poetic fame, the snow cover acts to insulate the plant against the subzero colds that can settle in for weeks in the thin upper air. In addition, deep snows provide water storage for the coming year. In a good winter, as far as huckleberries are concerned, the snow will fall thick and early. Soon it will cover the plant entirely, insulating the delicate tips of the newest growth from the cold and keeping them from drying out.

All winter, the plant waits and rests just like a bear in hibernation, systems all but shut down. Winds may blow, ice and hail storms rage night after night, temperatures plummet, and the huckleberry bushes lie protected under their snowpack.

After the winter solstice, bit by bit the sun increases its force, bringing highly changeable weather and precipitation: warm chinooks, possible avalanches, spring blizzards. Slowly the snowpack melts, and the first fingers of the huckleberry poke out above the receding crust.

This is some of the most capricious weather of the year—and some of the most dangerous for the huckleberry.

The danger comes because the protective snow covering, especially in a year when the snowpack is light, may melt away before winter harshness is over. Then the fragile bud-ends of the plant can be hit by cold, ice or hail, destroying the new growth. Worst of all, a false spring can lure the exposed plant into budding too early.

So far as anyone knows and most people assume, the prime cause of poor huckleberry crops is exactly this: a late frost. Activated by the warm sun, the huckleberry bush may bud and even blossom, only to be frozen by an all-too-common cold spell in May or June.

Perhaps surprisingly, most do survive the harshness of winter and caprice of spring, protected by snow or their own evolved hardiness. And now the huckleberry, being a precocious plant, will bloom before or at the same time as its leaves come forth.

Blossoms appear on last year's growth, usually singly along the woody stems that have emerged since the plant reactivated in the spring. The flower of the huckleberry is white or tinged with pink and shaped like a Chinese lantern with its petals fused into a single rounded bell. Inside the flower are two striking orange anthers with horns that can be seen with a 10-power hand lens. Very sexy, no doubt, among the huckleberry community.

The blooms are small and delicate, and stay on the plant so briefly that many berriers never see a huckleberry flower. Since the blossom shape is the prime characteristic used to distinguish among different species of Vaccinium (especially membranaceum and globulare),the briefness of its appearance further limits that effort. More critically for the huckleberry, the short flowering time means that fertilization must take place quickly if it is to take place at all.

The exact mechanism of pollination is not known, but it is assumed that bees and perhaps ants perform most of this work. Here is another potential pitfall for the berry-in-waiting: a cloudy spring—not uncommon in the Northern Rockies and Cascades—can effectively cancel the bees' pollen-carrying plans. The bees need a certain amount of sunlight, because their navigation depends on a visual memory of the path they take to their food supply. Without that light, bee activity is limited and the huckleberry blooms—orange anthers or no orange anthers—can go unfertilized.

Once the blossoms are fertilized, a tiny green fruit begins to swell at the base of the flower, which quickly drops to the ground. But danger is not past for the prospective berry. There is still the possibility of severe frost or drought causing the plant to drop unripe fruit. With luck, though, it will only be a matter of time and moisture and sun before that little green nub ripens, gradually deepening in color to red and finally purple, blue or glossy black.

At which point the huckleberry hunters converge on the hillsides once more, to reap the harvest of the Vaccinium's long year.

Bears & Berries

Many huckleberry secrets
Are known to the bears
Who deeply resent
Being caught unawares

—Mary Burke

Ursus and Others

Human beings are not the only ones who prize the huckleberry. Ripe huckleberries, which in their profusion can blanket whole square miles of mountain terrain, are a major food source for animals as well. Deer, birds, rodents and insects all dine off berries. Ptarmigan and grouse have easy picking on shorter shrubs, and even packhorses have been known to slurp them up when given the opportunity. In a drought year, one outfitter remembered seeing yellowjackets drinking from the ripe fruit. Poodles and Labrador retrievers may have to be taught to take their share from bushes rather than buckets, but otherwise will enjoy huckleberries along with their owners.

But the creature most often associated with huckleberries is the bear. As the Kootenai have known and celebrated for generations in the Grizzly Bear Dance, the berry is crucial to the bears' diet. Stories from the berrypatch all carry the edge of surprise, the awareness that at any moment a bear might appear to enforce her or his territorial claim.

Despite the obvious competition, outright conflict over huckleberries is highly unlikely and equally unnecessary. Berriers can improve their standing in the woods—and their understanding of huckleberries—by a look at the life and times of bears.

Black and Grizzly

The mystique of the wild huckleberry is great, but the mystique of the bear is greater. Some say this is because bears are capable of killing a person, but moose are also powerful and have been known to attack with serious injury or death. This lends credence to the second explanation, which is that the bear is the closest kin we have in the wilderness. The power of the bear—either in physical fact or in metaphysical myth—has long been acknowledged by native people, who frequently found guardian spirits in the form of a bear.

It is roughly accurate to say that wherever there are berries there are bears, although the reverse is not necessarily true. In any given season, fully thousands of people will converge on huckleberry patches along with the bears. Surprisingly enough, encounters are very rare, incidents of attack or injury almost unknown, and there have been no reported deaths of berrypickers.

Contrary to popular fears, the chance of suffering harm from a bear is just this side of infinitesimal. Bear expert Charles Jonkel points out that in any given year, between 100 and 300 people will be killed by lightning; 5,000 children will be killed by their parents; 50,000 people will die in automobile accidents; and 360,000 people will die from smoking cigarettes. And bear-related deaths? An average of one—every other year.

Statistically, then, one is in far graver danger driving to the trailhead than picking in the berry field. However slight, the risk is still present and can be lessened by an awareness of the bear's needs and patterns.

Bear Food Habits

To understand what huckleberries mean to bears, one must consider the animal's life and circumstances in general, and its diet in particular.

To a degree that human beings will have to stretch to imagine, bears have an enormous food dependency. They are huge creatures who move around a lot—when they move at all. Remember that bears sleep for five to seven months a year, during which time they are still consuming food energy. That leaves the other seven to five months in which the bears must eat enough to sustain them through both the active period and hibernation.

As a result, an adult animal needs to find and consume 80 to 90 pounds of food per day. As Jonkel says, "No wonder bears are crabby as a bear! They're busy enough surviving, without being troubled by having to run away from backpackers and berrypickers."

Vegetarian Bears

Again contrary to popular myth, bears—both grizzly and black—eat very little meat. It is estimated that more than 90 percent of the bears' diet comes from roots, grasses, leaves and fruit. Furthermore, what meat they do eat is often the carrion of animals that died of natural causes, or the remains of human-killed game.

As with huckleberries, the remoteness of the subject and difficulty of research so far have prevented unraveling the secrets of the bear. Still, several dedicated researchers have looked at the critical issue of bear food, particularly berries.

Research methods, it should be pointed out for the uninitiated, do not involve asking the bear what it likes or watching which foods disappear from the forest. Instead scientists use a sophisticated analysis of what they call scat, which is what everyone else steps around on the trail. Results show that the bear's diet varies dramatically through the year as different foods become available. While regaining strength during the critical period right after hibernation, April to May, bears clean up the carcasses of winter-killed animals, consuming much of the meat they will have that year. Into July the diet consists mostly of leaves and grasses, but as soon as berries ripen these become the food of choice.

In volume, well over half the bears' food comes from grasses and leafy matter, a favorite being the cow parsnip. Next are berries, comprising up to a third of the diet; huckleberries alone will account for 15 percent of the bear's annual intake. The smallest proportion is from mammals, with occasional ant or termite feasts providing a snack.

But these volume measures are estimates, and do not assess the importance of each food group to the bears' survival. While only 25 to 30 percent in volume, the berries' nutrient and energy content makes them more than a quarter or a third of the bears' total sustenance. Furthermore, to see berries from the bears' point of view, it helps to look at the time of year in which they become an important food source.

Huckleberries ripen late in the season, a high-energy dinner that supplies the largest part of the diet in the last months before hibernation. Often berries are eaten in a final feeding frenzy before the bear finally grows lethargic and beds down—not unlike other Americans around Thanksgiving time. Thus the berries are a critical element in the fattening process that will allow the animal to survive its period of hibernation.

The relationship between bears and berries is intense in the so-called "Northern Ecosystem" of the Montana Rockies, but it is not exclusive to huckleberry country. Wherever berries are found in Canada, Alaska and the Arctic, they constitute a significant element in the diet of the local bears. The aptly-named "bearberry" or kinnickinnick, cousin to Vaccinium in the heath family, is extremely important since it lasts through the winter and is available for the bears when they first emerge from the den. Polar bears eat a merged species of cranberry and kinnickinnick.

More than any other type of berry, huckleberries are integral to the survival of both grizzly and black bears. It seems no accident that the boundaries of prime grizzly habitat in the United States closely parallel the boundaries of prime huckleberry habitat. Bears who live in ecosystems without huckleberries fill in their diet with other foods. Yellowstone Park bears replace huckleberries with salmon and the nut of the whitebark pine; some think that this increased meat hunting creates Yellowstone's more aggressive bear population.

Individual bears may establish an ongoing relationship with a particular old burn site, and return to the same area year after year to feed on the berries there. In this sense the bears really do mark "their" patches, and will be rightfully troubled to find them picked clean by other hands.

One factor that helps avert seemingly inevitable conflict is that human beings generally pick their berries as low as possible in order to avoid extra climbing, while bears tend to seek huckleberries at higher elevations. Apart from the greater consistency of high-altitude crops from year to year, it also may be that the bears prefer the sweeter, more nutritious berry produced by greater ultraviolet radiation in the upper elevations.

One other aspect of bear-berry dynamics illustrates the fine match between bears and their chosen late-season food. Bears have an extraordinarily keen sense of smell—200 times as powerful as human beings'—and huckleberries have an extraordinarily powerful scent.

Jonkel explains it this way: bears are as sensitive to all odors as human beings are to the smell of skunk. In exactly the same way that a person driving in a closed automobile at 60 miles per hour can tell where a skunk died a week ago, a bear can cross the two-day-old trail of a human being and know not only that person's sex but probably have a pretty good idea of what she or he had for breakfast.

Combining this sensitivity with the powerful scent of a huckleberry patch—which most humans can pick up under the right conditions—it is easy to see how a bear can spot a good berry patch several miles away. These sensibilities are not always applied to huckleberries, of course. A woman who went huckleberrying left her car window open a crack to let the heat escape, returning only to find the whole window torn out by a bear. What she hadn't noticed, and what the bear could not fail to notice, was the scent of the garbage she had hauled to the dump a full week earlier.

Coexistence

Given the perpetual food crisis that is a way of life for the bear, and given the fragile foothold the bears maintain in their forest habitat, huckleberries should always be considered in context of bears' needs. As a matter of safety, if not courtesy or a sense of ecological balance, it behooves a berrier to consider whose larder she or he might be raiding.

Many people spend great portions of their lives in bear and berry country and never even sight an animal, probably because a bear is quick to catch human scent and disappear long before the person's relatively dim senses even register the bear's presence. The few who do encounter a bear should be prepared to react thoughtfully and prudently, with due respect for the bear's needs and habits. There are right ways and wrong ways to act, and consequences appropriate to each. Two stories illustrate this point.

One time a young girl was out berrying alone when she came across a full-grown grizzly. She saw the bear just as the bear saw her, and "I sneaked off in one direction while he sneaked off in the other," she said. This is the right approach, and an agreeable outcome.

Then there is the case of the man who came upon a small bear feasting in a berrypatch. The black bear was having the time of its young life with those huckleberries, munching away without a care or a nose to the world. Because the animal was small and not a grizzly, the man foolishly assumed he had nothing to fear, and decided to "have a little fun." So he sneaked up behind the creature, poked it in the back with a stick and yelled "BOO!"

Now this sort of person should not be allowed to walk around loose in town, let alone the woods, but sometimes justice plays out without human assistance. This was one of those times. Justice did not arrive in the form of an angry sow, though that possibility would have been obvious to all but an idiot; and neither was justice in the form of excess fury from a small beast whose survival has been threatened, though that too should have been considered at the outset. No, this time justice came in a split-second reaction that surely surprised the animal as much as his assailant. Turning to the direction of the poke, justice came from deep inside that bear's belly, up through its throat and out its mouth, showering the man with the vilest mix of half-digested huckleberries and bodily fluids of which a small bear would be capable.

In a final sweep of universal justice, the man never enjoyed a huckleberry again as long as he lived.

So far the superabundance of huckleberries (generally, if not in any one location) has kept the bears adequately supplied, and conflict kept to a minimum largely thanks to the bears' keen senses and graciousness in sharing their summer food. With public awareness and sensitivity, there is no reason that Homo sapiens and Ursus arctos cannot continue to coexist in the berrypatch as they did in the early days. As one woman remembers her mother saying many years ago, "God made those huckleberries for us both."

A Different Perspective

Rhoda Cook outfitted in the Bob Marshall Wilderness for 27 years. In all that time she saw only three grizzly bears, none of whom caused her a problem. Feeling a certain kinship towards bears, they were not her preferred hunting game, although she did take a dozen or so black bears in her guiding career. For one thing, she never believed she could make a decent pie without beargrease. Her husband, however, was a gentle soul who refused to kill game of any sort, leaving that task to Rhoda; furthermore, he didn't care for bear meat in the slightest. "I always know if it's bear, it just gets bigger and bigger in my mouth," he said.

Once he caught a glimpse of one of Rhoda's bears in the shed where she hung game. Echoing the sentiments of many who have seen a skinned-out bear, he shuddered and said, "How can you do that? It looks just like a man."

Rhoda considered that. "I don't know," she said. "I've never seen a man strung up."

Huckleberry Hunting

Sometimes, having had a surfeit of human society and gossip, and worn out all my village friends, I rambled still farther westward than I habitually dwell, into yet more unfrequented parts of the town, "to fresh woods and pastures new," or while the sun was setting, made my supper of huckleberries and blueberries on Fair Haven Hill, and laid up a store for several days.

The fruits do yot yield their true flavor to the purchaser of them, nor to him who raises them for the market. There is but one way to obtain it, yet few take that way. If you would know the flavor of huckleberries, ask the cowboy or the partridge. It is a vulgar error to suppose that you have tasted huckleberries who never plucked them.

A huckleberry never reaches Boston; they have not been known there since they grew on her three hills. The ambrosial and essential part of the fruit is lost with the bloom which is rubbed off in the market cart, and they become mere provender. As long as Eternal Justice reigns, not one innocent huckleberry can be transported thither from the country's hills.

—Henry David Thoreau, *Walden*

Fishin' is different than catchin'.
—Anon.

Berrypatch Philosophy

Anyone looking for a map will be disappointed here, finding no trace of routes or drainages that might yield a crop of plentiful purple. No, now is not the time to be giving away secrets carefully preserved for generations. If as a result the new berrier takes a little longer to land the first patch, he or she will be all the prouder for it. Furthermore, to presume the berry places that were good in the last year or 10 or even 50 are going to be good by the time anyone reads this is presumptuous in the highest degree. Word travels fast in huckleberry country, and the mere mention of a good spot could be borne by the wind and cause the frost to fall on little pink flowers. It's not worth the chance, and besides—the joy is in the hunt.

The most important thing to bring to a huckleberry expedition is the right attitude. Those who arrive cheerful and open to any adventure will enjoy their day, and likely their berries, more. The unpredictable nature of huckleberries is actually a reflection of wildness itself. When you go into the woods, you can never be sure exactly what you'll find, but taken in the right spirit, whatever you find will be a treasure.

And it is possible to discover treasures while berrypicking. One woman happened to look up from her picking, and see a silver baby cup hanging in a tree overhead. She recognized the name engraved on the cup and was able to return the keepsake to its owner, whose mother had left it there years ago. And somewhere out there are still those 21 silver dollars buried by mean old Carlson's kids back in the 1940s. The point is to be alive and aware and ready: for stories, for treasures, for anything. Nothing kills a huckleberry trip faster than a businesslike quota.

No, huckleberrying is not a good time to be preoccupied with mundane events. Rather it is an ideal place to think through the more weighty matters of life. As Rhoda Cook said,

"It's therapy. I've buried three husbands, two sisters, both my parents and a daughter, and I can tell you I've done a lot of thinking out there. Picking huckleberries gets you away from the telephone and away from the typewriter, out in Nature where you can connect with your values. I always come out remembering 'nothing's going to happen today, God, that you and I can't handle together…'"

At times like that, huckleberrying may be best done alone so one can think without interruption. But sometimes it's nice to have company, or to bring along someone who might like to do some quiet thinking of his or her own. There is something salutary in the combination of fresh air, adventure and the honest work of picking huckleberries. Most world problems have been solved at least once in a berry-patch, and many personal problems as well.

If you do bring company, choose carefully. You'll want to consider picking styles and philosophies, because as we'll see, not everyone gets along in the huckleberry patch.

Commercial Picking

The going price of a gallon of huckleberries may at first make commercial picking an appealing prospect. But before quitting your job and buying a huckleberry hat, remember that the work is difficult and uncertain, and for the price you get—as one picker reminded her California buyer—you've already taken your life into your hands.

One man was about to buy a gallon of huckleberries when he heard the price. He stopped short, snorted, "At that rate, I'll go pick them myself!" and off he went. He returned that very afternoon looking a little worse for the wear, and handed over the money. "You're not charging enough," he said.

Those who are ready for whatever the wilderness dishes out, and who like the freedom it offers, find it is possible to make useful money picking huckleberries. One fellow decided to pick and sell enough berries to build a pig barn. When it turned out to be easy picking that year, in no time he had the pig barn up, was working on feed for the winter and figuring out how to get by on pigs and huckleberries from then on.

A little extra money can come in handy for anyone, as it did for the new father who earned diaper money selling a few huckleberries. The canning company buyer shook her head; "I'd never bought huckleberries from anybody wearing a tie before."

The Habitat

Where to look? Some like to say that huckleberries grow best where they're hardest to get to, but the truth is that once you're in huckleberry country, it's reasonable to look for them almost anywhere. You'll find huckleberries where they get what they need; a detailed survey of this is found in the fourth chapter. For the purposes of scouting while driving or hiking, these guidelines will help you spot the most likely places:

- Elevation above 3,500'
- Open or partially shaded slopes
- Burn areas older than 15 years
- Relatively moist areas (indicated by presence of other shrubs and vegetation; not dry grassland)

Perhaps most importantly, especially for the novice, choose a trail or road in a drainage that you simply feel like exploring: one that looks especially inviting to you, has an amusing name or that you've never been in. That way you will enjoy your outing regardless of how many huckleberries you find. Keep your curiosity close at hand while walking or driving. If you see something that looks interesting—foliage, berries, anything—stop and investigate. You never know what you'll discover.

Finally, if the crop seems terribly scanty and you become discouraged, remember this advice: "You'll find berries every year if you know where to look—

"...and keep looking."

The Berry

How do you know when you've found a huckleberry? Good question. Primary characteristics are:

- Purple, deep blue or red color, with or without a dusting of "bloom"
- Pea-sized berry with smooth skin, seeds no bigger than flakes of pepper
- Bushes knee- to chest-high, green leaves occasionally tinged with red

If the berries, and the bush they came from, look like the picture on the cover of this book, chances are very good that what you have are huckleberries. Most of the other shiny purple berries in the woods are perfectly edible, but if you pick something else you'll be disappointed or worse, depending on how wrong you were. Most cases of mistaken identity involve serviceberries, which grow on taller bushes, are less juicy and more coarse in texture, and when squeezed do not stain the fingers as brilliantly as the huckleberry.

To be absolutely certain that what you have in your hand is what you want, ask an expert or look in the field guide that you thought to bring along on your expedition. No expert? No field guide? Not sure? Don't start with a taste test. Instead pick a handful or a hatful and consult someone who knows. Once you have a positive identification, memorize the leaf and berry structure or press a sample for future reference.

The Season

Telling someone when to pick is almost as unlikely as someone telling you where. It varies from year to year, location to location, and elevation to elevation. July 25 is a good date to remember, but not to count on. In an early summer, the first low-elevation harvest might come as early as June; in a late season it may be well into August before a single berry darkens in the higher elevations. Indian Summer berries of late September or even October can be the sweetest of all, if you can find them. Of course there are the stories of people picking huckleberries through snow on the way to hunting camp. Take nothing for granted.

The Harvest

Once you've found a patch to your liking, it's time to pick huckleberries. Simple as it may seem, at this point the berrier is faced with a very important decision. Huckleberries are relatively small and spread out over the bushes, which in turn may extend for several yards, acres or square miles. Getting the berries off those bushes is the first issue, and no small one.

Even more than what to call a huckleberry, the proper method of picking is probably the source of more arguments than other issue relating to huckleberries. The scientific controversy over nomenclature is but a spring breeze compared to the tempests waged over how to pick berries.

Huckleberry argument turns violent...

An argument about the proper way to pick huckleberries led to an assault at the Veterans of Foreign Wars Club in Whitefish, according to the Whitefish Police Department.

A Whitefish man was discussing huckleberry picking when one man knocked him to the ground and another kicked him repeatedly in the face, according to Detective Dan Voelker.

Two other Whitefish residents were arrested and charged with misdemeanor assault. They were released on $200 bond.

—Whitefish [Montana] Pilot, *August 12, 1987*

If it followed typical lines, the Whitefish argument was probably another round in the ongoing dispute between "beaters and pickers." Three methods are widely used in huckleberrying, and proponents of each are unlikely to admit that any other is even worthy of discussion.

• **Pickers.** One device used to remove berries from bushes is a huckleberry "rake." Like all huckleberry equipment, these are usually handmade according to family design and preference, and no two will be exactly the same. The basic principle is that of a comb. A set of metal teeth, spaced strategically to trap berries, is attached to a coffee can or other receptacle. The whole thing is outfitted with a handle, and all the picker has to do is reach into a fruiting bush, get the comb behind a string of berries, and pull; the berries fall into the can.

• **Beaters.** The second main type of equipment used for berrypicking, again homemade and of varying design, is that of "bags and beaters." Here some kind of paddle is used to hit the bushes, causing the huckleberries to fall off and be caught in a receptacle below—either a piece of material laid on the ground, or a shallow bag with a very large opening that is held under the berries.

• **Fingers.** The oldest picking method known is also the simplest. By hand, berries are gathered a few at a time, and then either dropped directly into one's pail or allowed to collect in the palm until a whole handful can be deposited at once.

The controversy? Each method has its pros and cons. Hand picking is criticized for being slow, but both pickers and beaters collect a substantial number of leaves and twigs along with the berries, which then must be removed. Time spent in cleaning can take much of the time saved over hand picking.

On the other hand—and this is where the argument really heats up—many people believe that the use of either pickers or beaters damages the huckleberry plant. With fewer leaves, the bush has less available area for photosynthesis, and the stem ends that can be broken off are the new growth from which next year's crop is expected. Pickers blame this on beaters and beaters blame it on pickers, but both claim any damage to the plant is simply pruning and jointly criticize hand pickers for depriving the plant of this beneficial process.

There is yet one other method of picking, largely unheard-of now but reported in the past as practiced by the native people. Children would go into the berrypatch and collect whole stems of berries, bringing them back to older women who would sit and pick the berries off the stems and discard the brush. Again, this was thought to be a beneficial pruning.

Who is right? The unpredictable nature of the huckleberry crop means that no one knows exactly why a patch produces well when it does, or poorly when it doesn't. Was it a rough session last year with an overenthusiastic bush-beater, or an isolated frost pocket one otherwise warm night in May?

Common sense suggests that any practice that damages the huckleberry bush should be avoided. Pickers, beaters and stripping brush may not have noticeably damaged the huckleberry crop in the past, but as commercial and recreational use of the berry increases and the old burn areas are reforested, this may change. Until more is known, a conservative approach seems warranted; whenever possible, berries should be picked by hand. If other methods are employed, care should be taken to do as little damage as possible to the bush itself.

There is one other difference of opinion that may not matter to the huckleberry but matters a great deal to the individual berrier. That is: to munch or not to munch. Many devoted pickers will spend entire days in the berrypatch and never mouth a single morsel, as a matter of principle. Other equally devoted berriers eat as many as they can while picking, making the most of their presence and the least of their load on the way home.

Really Good Picking

No matter what method is used, the picking of huckleberries is going to seem slow, at least compared to the eating. Harvesting huckleberries is hard work, not without risks, and certainly not for everybody. There was a family from eastern Montana who drove many hours to try huckleberrying. Their host brought them to a prime patch, which he left to the family while he moved to some bushes nearby. Thirty minutes later the father called to Tex: "Do you call this good picking?"

"Yes," Tex said happily.

"We're going home," the father called back. They did.

The best hand-pickers say they can fill a gallon in less than two hours if it's "very good picking." More commonly, a gallon will take three to four hours to pick. Using pickers or beaters, one can bring home 10 to 20 gallons in a day.

Other measures can be important, since berrypicking is a physically demanding activity. One woman says she can count the passing years by how long it takes before her back hurts; once it was three days, but now it takes only a day and a half.

Picking Tips

• Pick uphill, from bottom to top of a slope; it's easier to see the berries underneath the leaves.

• Vary your position frequently. Stand, sit or bend from side to side; kneeling can protect a rheumatic back.

• Whistle while you pick (especially children with purple lips); the bucket will fill faster.

• Once you find a generally good band of berries, hunt new patches at the same elevation along the mountainside.

• Never laugh at someone else's berry-blotched bottom before you've taken a look at your own pants.

Collecting

Except for those who simply pick and then eat a handful at a time along the trail, berriers need some kind of collection system to bring their huckleberries home. The most popular article for this purpose is a pail of some kind. Recreational grade is often a three-pound coffee can with an improvised handle. More elaborate inventions include cans or buckets with lids (preserving the day's catch in the event of a stumble in the mountain brush). Commercial harvesters often devise elaborate belt-, back- or chest-packs to carry large quantities of berries back to camp.

When using a bucket, it is a good idea to line the bottom. The large, soft leaves of the thimbleberry can be used for this; paper towels serve the same purpose. The problem with large berry cans, even five-gallon size, is that a full load of berries can crush the bottom layer. Any leaves and twigs will be very difficult to remove from the mush. For premium-quality berries, collect them in the smallest practical quantities, and store them in the shallowest possible layers. The flats of beer cases make good containers for transporting berries home.

Keeping Fresh

The importance of freshness is not to be underestimated. As with any food, degeneration sets in as soon as the berry is picked; when time combines with heat or pressure, the result will be a loss of quality in your pie, jam or other mouthful. The less they are handled, the better.

Keep berries cool and out of the sun. A mountain-cold creek can be used to bathe a collection bucket, but putting the huckleberries directly into water is not advised since their flavor can be quickly diluted.

Cleaning

Cleaning berries—removing any non-huckleberry material—can be a time-consuming process, depending on the dampness of the berries and the amount of twigs and leaves that came along. Hand-picked berries usually require minimal cleaning, if any, but those who use beaters and pickers may need an elaborate system to separate berries from dross.

Apart from picking the debris out by hand, a procedure not recommended except for small quantities of berries or saints-in-training, the simplest method for cleaning uses a board and a blanket. The blanket is wetted and smoothed onto a board set at about a 45° angle. Berries spilled onto the top of the board are allowed to roll down the blanket into a waiting receptacle, leaving the leaves and twigs stuck to the blanket.

Other common cleaning methods also use the principle of a sloping surface. Often it will be a trough 10 to 12 feet high and 20 feet long. The trough may be lined with either screen or fabric to collect the leaves, or parallel lines of wire or string that allow leaves but not berries to drop to the ground below. A further enhancement is to build in a "bounce plate," a drop-off to shake loose otherwise stubborn leaves.

Even hand-picked clean berries, if destined for a decorative dessert, can be polished: put a dry turkish towel on a board at about a 30° slope; run the berries down the towel a few at a time.

Basic Woods Wisdom

The dangers of huckleberrying, compared to rock climbing or skydiving, are scant. However, it is to be remembered that berries are found in wilderness, and wilderness must always be approached with respect. Some considerations:

• **Weather:** While most berrying takes place in the mild summer and early fall, there is always the chance of an unexpected storm or change in temperature. Be prepared! Bring raingear and some warm clothes, and a complete change if your trunk is large enough. Only under severe circumstances will this be a survival issue, but when something does happen it can easily feel like life and death. Keep an eye on the sky while your hand is in the bush. If it looks like rain, consider your options; those clouds could also contain hail, or in freak instances and high elevations, snow. Don't assume you will be warm in 20 minutes just because you are warm now.

• **Navigation:** When concentrating on filling a berry pail, wading in and out of deep brush off the trail, it is not at all difficult to become disoriented. Always bring along a map of the area. When you settle in for serious picking, choose a landmark to guide you back to the trail and check it frequently to make sure you haven't wandered off into unknown territory. If the difference between Douglas fir and ponderosa pine eludes you, hang a bright sweater or other visible item on your chosen mark.

• **Bears:** Considering all the trouble one's own foolish enthusiasm can generate—not to mention the unfortunate potential for trouble of the two-legged kind—bears really comprise a small portion of the danger in the woods. However, because berriers are in direct competition for the bears' cherished (and essential) food supply, do observe these precautions:

1. Follow all established guidelines for behavior in bear country. Make noise, avoid cooking and eating odiferous foods, and keep your eyes and ears open. Most berriers have no problem being noticed. "I just hang a couple buckets on my butt and no bear will come within 40 miles," one laughs.

2. If you see a bear, or suspect one nearby, don't challenge it! Retreat as if a blizzard had just struck—out of a spirit of self-preservation, if not simple courtesy. You are inviting yourself into the bear's

home, and while few will attack without cause, the bear gets to decide what is cause and what is not. Don't risk it!

• **Survival Kit:** Almost any guide to hunting, fishing, or camping will include suggestions for a small "survival kit." Basic ingredients include matches, candles, string, first aid gear and emergency food rations, space blanket, saw, and other items that truly can help save a life—your own or someone else's—under certain conditions. A good kit can pack all of this and more in a space smaller than a canteen and still weigh under a pound. It's easy, it's sensible, and simply having it along makes any adventure feel more adventurous.

A Way of Life

Just as huckleberries don't suit everyone, huckleberry hunting is not a sport for everyone. But those who like it may find they like it a great deal. For many people, huckleberry hunting is a much-anticipated and beloved tradition. The berrypatch is one place where anyone can shed pretense and simply be outdoors enjoying what the natural world has to offer.

Huckleberry traditions are passed from generation to generation, and there are those who figure it's in their blood. The women in Jan Dunbar's family are carrying on an irreplaceable tradition when they head to the berrypatch each year:

"My grandmother was a character, a superb cook and a delight to be with. My mother, Beth Romney, learned about huckleberries from her and was never the same again. She would disappear with her lard bucket, wearing her Huckleberry Finn outfit, and spend hours bent over or sitting in the patches, totally happy. Wherever her particular heaven is, I know there are huckleberry bushes all around.

"The last photo I have of her is a small colored one showing her with my daughter, Dusty, picking like mad. Mother was 84. It was two years before she died. Dusty had to hoist her up the bank from the road by pulling her under the arms. The old fanatic look is on her little face, and it is a real treasure to me, those two precious girls out huckleberry-ing together."

How to Eat Huckleberries

The Gatherers

The word gets out like Sutter's Mill.
They're on! It's happening, like gold in the sluicebox,
the huckleberries are ripe. Now.
I can smell 'em.

> The ritual of the gatherers.
> Tin lard buckets.
> The grandmothers. The great-grandmothers.

It is at dawn we go, the daughter and I.
Thirty miles away at sixty miles an hour.
I can smell 'em.

> He thinks we're crazy
> as he ties the leader to the line
> here on the dry river of his study.
> A good knot. Might hold a three-pounder.

Crazy, yes. Mosquitoes, horse flies, ants,
bears, and what do you have?
A stiff back, purple fingers,
minuscule berries for scrawny pies,
so why do we do it?

> His eyes ask without looking up
> from his fly box.
> He's right, and we know it.
> I don't know why we go.

Later, deep in the knee joint, there's a message.
We do this now.
Our mother did it.
Our grandmother did it.
They and their mothers gathered and knelt,
pushing aside crinkled leaves for these small treasures
since forever, and before that.

We go.

> It's like fishing, I tell him,
> but I see he doesn't understand.
> The trout he puts back to catch again,
> but the pie he'll only eat.

—Jan Dunbar

Everybody Likes Huckleberries

It's not true. Everybody does not like huckleberries, even though most who don't are smart enough not to make a lot of noise about it. Neither is it true that you either love them or hate them. Some people like huckleberries in jam or candy, but not baked into muffins or pancakes; some like them fresh and only fresh. Others like huckleberries to the tune of one piece of pie a year.

Some, on their first taste, will say it's a little tart... and quickly follow with, but you can sure taste the flavor! Then there are those who like huckleberries without knowing it, like the fellow who always said he couldn't even stand the smell of huckleberries—until the day he was presented with a bottle of huckleberry wine and told it was fruit juice. He finished the whole pint without complaint.

Of course, there will always be the crusty types who don't like something just for the sake of not liking it. One such person caused a stir by announcing to a crowd of Montanans in the middle of huckleberry country that he had no use for huckleberries. After a deadly pause someone else warned, "Not liking huckleberries is like not liking apple pie!"

The old iconoclast looked pleased. "I don't like apple pie neither," he said.

Eating Huckleberries

Although there are many different styles of picking and preparing huckleberries, and as many reasons as people for doing so, there is only one correct fate for a huckleberry. Once picked, it is to be eaten. Preferably it is to be eaten by a sentient creature who understands the extraordinary nature of the fruit she or he is consuming—its history, character and value—but eating is proper in every regard so long as it is accompanied by enjoyment. One should eat huckleberries with the same relish, the same zest and freshness, the thirst for the unexpected, that characterize a good trip into mountains.

So there is no wrong way to prepare or eat a huckleberry. It is possible, however, to make a painful mistake. Remember Aunt Rose, who picked huckleberries all day with her family, came home and baked a big beautiful pie. It wasn't until the first slice that she realized she had mistakenly exchanged the salt for the sugar container. Thee is warned.

Please note: Like huckleberries and anything else associated with wilderness, the recipes in this chapter come without guarantees. They are included as a sampling of the many ways berries have been prepared and used over the years. Eat—as you pick—at your own risk.

Eating Huckleberries at Home

—————— Pies & Cobblers ——————

Huckleberry Cream Pie

3/4 c. sugar 2 1/2 T. cornstarch 1/2 t. salt 2/3 c. water 1 c. berries	Combine in saucepan. Bring to a boil and cook until thick and clear.
2 T. butter	Add butter. Cool this mixture.
2 c. berries	Fold in berries. Cool one hour.
1 c. cream, sweetened and with vanilla graham cracker crust	Beat cream and put half of it on the pie crust. Pour on filling and top with remaining cream. Chill until firm. The bottom cream layer in this pie keeps the berry juice from soaking the crust.

Governor's Pie

This is a favorite of Ted Schwinden, whose two terms as governor of Montana began in 1980.

Filling:
  4 c. huckleberries
  1 c. sugar
  3 T. cornstarch
  1/4 t. cinnamon
  1/4 t. nutmeg
  3 T. lemon juice
  1/3 c. pecans, chopped
    pinch salt

Mix together.

(Continued on next page.)

Pie crust:
1½ c. Crisco shorten-
    ing
1 t. salt
3 c. flour
1 c. cold water
  half and half
  sugar

Combine Crisco, salt and flour, and mix well. Stir in the water. Divide dough and roll out two crusts. Place one crust in the bottom of a 9-inch pie pan. Pour in pie filling. Cover with remaining crust and seal edges. Brush top crust with half and half, and sprinkle a little sugar on it. Bake at 350° for ½ hour, then 325° for 15 to 10 minutes.

Kim Williams' Husband's Montana Huckleberry Pie

4 c. fresh or frozen
    huckleberries
¾ c. sugar (more if
    berries are very
    tart)
3 T. flour
  pastry for
  double-crust
  9-inch pie

Mix flour and sugar with berries. Set aside. Roll out a little more than half the pastry dough to ⅛-inch thickness and fit into a 9-inch pie pan (Kim said, "do try whole wheat pastry flour"). Pour the berries into the pie shell. Roll remaining pastry a little thinner than for the bottom crust. Prick with your favorite design. Brush the edge of the bottom crust with cold water and place the upper crust on pie. Press crusts together at the rim and trim off the excess dough. Flute the edge. Bake in a preheated 450° oven for 10 minutes. Reduce the heat to 350° and bake about 40 minutes.
Variation: Add 1 T. vinegar to the pie filling.

Huckleberry Dumplings

2½ c. huckleberries
  dash salt
⅓ c. sugar
1 c. water
1 T. lemon juice

Measure ingredients into two-quart saucepan. Bring to boil. Cover. Simmer 5 minutes.

1 c. flour	Mix like shortbread.
2 t. baking powder	
2 T. sugar	Drop six dumplings into boiling mixture.
1/4 t. salt	Cover. Cook over low heat 10 minutes.
1 T. butter	
1/2 c. milk (stir in all at once)	

Grandmother Beatie's Fruit Cobbler and Whiskey Dip

Grandmother Beatie of Whiskey-dip fame was actually raised a devout Mormon. Born Helen Cordelia Clawson in New York, she was 12 when her mother left her father and took the four children to follow the Mormon movement to Nauvoo and Salt Lake City. Helen drove one of the wagons part of the way. The man who became Grandfather Beatie claimed that the word of wisdom forbidding drinking "came later" and was in his opinion a terrible mistake.

Cobbler:

2 c. flour

4 T. (rounded) butter, Crisco, or half of each

2 t. (rounded) baking powder

dash salt

milk

huckleberries

Sift together dry ingredients. Rub shortening into flour as for pastry (best if you use your fingers). Add enough whole milk to make the dough sticky. In a slightly oversized ovenproof pan or casserole, cover berries lightly with sugar and stir. Lightly drop crust mixture in spoonfuls over the fruit, pushing it against the sides to adhere and to completely cover top of fruit and to stick to the sides of the pan, "sealing" the fruit inside. (A deeper pan will prevent boiling over.) Bake in 350°-375° oven until brown and bubbly, and fruit is cooked. Serve warm under Whiskey Dip.

Whiskey Dip:

3 T. butter (level)

3 T. flour (level)

1 1/2 c. warm water

Melt butter and add flour as for a white sauce, stirring as it bubbles.

Add water. Whisk smooth over heat.

(Continued on next page.)

3 T. sugar	Add to mixture. Boil gently until clear.
pinch salt	Add water if too thick; sauce should be
$1/_2$ t. nutmeg	the consistency of thin gravy. Keep hot.

| 3 T. whiskey (or more to taste) | Whisk in whiskey just before serving. |

Preserves & Preserving

Spiced Huckleberry-Peach Jam

4 c. chopped or ground peaches (about 4 lbs.)	*To prepare fruit:* Sort and wash fully ripe peaches; peel and remove pits. Chop or grind peaches. Sort, wash and remove any stems from fresh huckleberries.
4 c. huckleberries (about 1 quart fresh berries)	
2 T. lemon juice	*To make jam:* Measure fruits into a kettle; add lemon juice and water. Cover, bring to a boil, and simmer for 10 min- utes, stirring occasionally. Add sugar and salt; stir well. Add spices tied in cheese- cloth. Boil rapidly, stirring constantly, to 9° over the boiling point of water, or until the mixture thickens. Remove from heat; take out spices. Skim. Fill and seal containers. Process 5 minutes in boiling water bath. Makes 6 or 7 half-pint jars.
$1/_2$ c. water	
$5^1/_2$ c. sugar	
$1/_2$ t. salt	
1 stick cinnamon	
$1/_2$ t. whole cloves	
$1/_4$ t. whole allspice	

Huckleberry Jam

4 c. berries	Prepare as for Huckleberry-Peach Jam
1 c. sugar	(above). This makes a preserve that is
4 T. cornstarch	more like pie filling than regulation jam.

Huckleberry Jelly

3 qts. (12 cups)
    huckleberries
1 large orange
    sugar

Cut orange into very thin slices; seed, but leave peel on. Cover with cold water and let stand in refrigerator overnight. Next day, wash berries, pick carefully, then mash and add orange slices. Heat slowly until crushed and all liquid is out. Strain, discard orange and skins. (Pulp may also be added to jam or used to make vinegar.) Measure out juice and an equal amount of sugar. Bring juice to boil, pour in sugar all at once, boil 10 minutes. Skim, pour into hot sterilized jars and seal with paraffin. Makes 10 6-ounce glasses.

Huckleberry Syrup

Mash a quantity of huckleberries with a potato masher, or whirl in blender for a few moments. Place in a cheesecloth or other fine strainer and let drip overnight.

3 c. huckleberry
    juice
7 c. sugar

Mix these amounts of huckleberry juice and sugar, or use similar proportions. Bring mixture to a boil.

$^1/_2$ bottle Certo

Add this amount (or proportionate amount) of Certo. Boil one minute.

Huckleberry-Raspberry Jam

1 qt. huckleberries
1 qt. red raspberries
7 c. sugar
$^1/_2$ bottle fruit pectin

Wash and crush fruits, combine, measure 4 cups. If necessary, top off with water to obtain 4 full cups. Add sugar, mix well. Heat to full rolling boil. Boil hard 1 minute. Stir constantly. Remove from heat and stir in pectin. Skim. Seal in hot, sterilized glasses. Makes 11 6-oz. glasses.

Frozen Huckleberries

Clean, rinse and freeze huckleberries the same day they are picked. Double-seal packages with glass containers and plastic, or two layers of plastic, because the pungent smell of the huckleberries can and will flavor everything else in the freezer.

Dried Huckleberries

Spread one layer of huckleberries onto cookie sheets and set in the sun on the hood of the car. Take in at night or before rain. Let dry until very light, the size of BBs: 5 to 6 days.

Can also be dried in a modern food dehydrator.

Pressed Huckleberries

Not for eating, but for when you want to use huckleberries in decorative crafts. Slightly dry branches with leaves and berries in sun or oven. Press between large telephone books.

— Pancakes, Muffins & Quick Breads —

Sour Cream Huckleberry Pancakes

1 c. flour	Sift dry ingredients together. Beat
3 t. baking powder	together egg, milk and sour cream. Pour
$^1/_4$ t. salt	milk mixture over dry ingredients, blend
1 egg	until smooth. Add melted butter, mix
1 c. milk	well, fold in huckleberries. Bake on hot
$^1/_4$ c. sour cream	griddle.
$^1/_2$ c. huckleberries	
2 T. melted butter	

Huckleberry Muffins

Two great secrets for making muffins with wild berries: one, flour the berries and, two, don't overmix the batter.

2 c. flour
2 T. sugar
$^1/_2$ t. salt
2 t. baking powder
1 scant t. baking soda

Sift dry ingredients together in round-bottomed bowl.

1 c. drained but damp, fresh or canned huckle-

Stirring from edge to center, add huckle-berries to the dry ingredients. Stir gently until they are all separated and coated with flour.

1 egg, slightly beaten
2 T. melted butter
$^3/_4$ c. buttermilk

Add to mixture. Stir just enough to dampen all ingredients. The mixture will be thick, more like sticky dough than batter.
Use an ice cream dipper or tablespoon and fill greased muffin tins half full. Bake at 400° for about 18 minutes. Makes about $1^1/_2$ dozen muffins.

Huckleberry Whole Wheat Muffins

1 egg
$^3/_4$ c. milk
$^1/_2$ c. vegetable oil
1 c. whole wheat flour
1 c. white flour
$^1/_3$ c. sugar
3 t. baking powder
1 t. salt
1 c. huckleberries

Heat oven to 400°.
Grease muffin tins or line with papers.
Beat egg, stir in milk and oil. Stir in dry ingredients until moist (batter will be lumpy). Fold in berries. Fill muffin cups $^3/_4$ full. Bake 20 minutes. Makes 12-14 muffins.

Huckleberry Bran Muffins

1/4 c. vegetable short-
    ening
1/4 c. brown sugar
    (packed)
2/3 c. milk
1 egg
1/4 c. maple syrup
1 c. flour
1 c. bran
1 T. baking powder
1/4 c. salt
2/3 c. huckleberries

Beat together shortening and sugar until light and fluffy. Add milk, egg and syrup, then add remaining dry ingredients, mixing until moistened. Add huckleberries. Grease muffin cups and fill 2/3 full. Bake 15 minutes at 400° or until wooden pick inserted in center comes out clean. Makes 12.

Huckleberry Fritters

3 T. sugar
1 c. commercial
    biscuit mix
1/3 c. milk
1 egg, beaten
1 c. huckleberries
    powdered sugar

Mix ingredients together and drop by spoonfuls into hot fat. When they are nicely browned, drain on paper towels or paper bag and sprinkle with powdered sugar.

Huckleberry Waffles

2 eggs
2 c. buttermilk
2 c. flour
2 t. baking powder
1 t. soda
1/2 t. salt
1/4 c. plus 2 T.
    shortening
1 c. huckleberries

Heat waffle iron. Beat eggs, beat in remaining ingredients with rotary beater until smooth. Pour batter onto center of hot waffle iron. Sprinkle 2 T. huckleberries over batter for each waffle as soon as it has been poured onto iron. Bake about 5 minutes or until steaming stops. Remove waffle carefully. Makes about eight 7-inch waffles.

Cakes, Candy and Desserts

Rocky Mountain Huckleberry Cake

2 c. flour
2 t. baking powder
$1/4$ t. salt
$1/8$ t. cinnamon
$1/8$ t. nutmeg
$1/2$ c. butter
1 c. sugar
2 eggs, separated
$1/2$ c. milk
2 c. fresh huckleber-
   ries, lightly
   floured

Sift flour with baking powder, salt and spices. In another bowl cream butter, sugar and egg yolks. Beat egg whites until stiff and set aside. Add flour mixture and milk to butter, sugar, egg yolk mixture. Stir in huckleberries. Fold in egg whites. Pour into greased 9-inch round cake pan. Bake in 350° oven about 45 minutes. Sprinkle lightly with powdered sugar. Serve warm.

Huckleberry Cake

1 c. sugar
1 c. sour cream
2 eggs
1 t. vanilla
2 c. flour
1 t. salt
1 t. baking powder
2 c. clean huckleberries

Mix sugar, sour cream, eggs and vanilla together. Add dry ingredients, mixing well. Gently fold in huckleberries. Bake at 350°, about 30 minutes. Serve with cream or ice cream.

Huckleberry Upside Down Cake

$1/4$ c. margarine
$1/2$ c. sugar
2 c. huckleberries
1 t. lemon peel
1 t. vanilla
1 $8^1/_2$-oz. pkg. Jiffy
   cake mix
$1/2$ c. heavy cream,
   whipped

Melt margarine in 8-inch square pan. Sprinkle sugar evenly over melted margarine. Mix huckleberries with lemon peel and vanilla and sprinkle over sugar. Prepare cake mix according to package directions. Spread over huckleberries. Bake in 375° oven 30 minutes or until cake is done. Let stand for 10 minutes. Turn onto platter. Serve warm with whipped cream.

Carrie's Cake

4 c. huckleberries
3-oz. pkg. Jello
1 c. sugar
3 c. miniature
    marshmallows
1 pkg. yellow cake
    mix

Grease a 9x13-inch pan. Spread berries on bottom of pan. Sprinkle with sugar and Jello. Top with the marshmallows. Prepare cake mix according to package directions. Spread batter over berries. Bake in 350° oven 50-55 minutes. Cool 5 minutes and then turn upside down onto platter. Serve with whipped cream or Cool Whip.

Mildred's Huckleberry Chocolates

Fondant:
$^1/_2$ c. soft butter
$^1/_4$ c. corn syrup
$2^1/_2$ c. powdered sugar
    huckleberry jam
    whole huckleberries

Mix butter and syrup in large bowl. Work in powdered sugar and knead with hands. Add huckleberry jam to taste. Reserve the whole berries. Chill.

Coating:

Test dipping chocolate to assure that it is compatible with mold and removes easily. Heat chocolate until just melted, not hot, taking care not to allow any water into the pot (chocolate will curdle). Coat molds with a solid layer of chocolate, using melon ball cutter, baby spoon, or other utensil to get a smooth, evenly thick shell. Chill until hard. Hold mold up to light; reinforce any thin areas with more chocolate. Fill each mold with fondant and one or two whole huckleberries. Cover with final layer of chocolate, taking care to seal edges around the fondant.

Huckleberry Dessert

24 marshmallows
$^1/_2$ c. milk
1 t. vanilla
1 c. cream, whipped

In double boiler combine milk and marshmallows. Cook, stirring, until marshmallows are melted. Add vanilla. Let cool then fold in whipped cream. Set aside.

18 graham crackers, crushed
$^1/_4$ c. butter
$^1/_4$ c. sugar
1 qt. fresh or frozen huckleberries

Mix together graham crackers, butter and sugar. In 9x13-inch pan layer half the crumbs, half the marshmallow mixture, all huckleberries, remainder of marshmallow mixture, and top with remainder of crumbs. Refrigerate overnight or several hours.

Other Treats

The two recipes that follow are basically old-fashioned home wine recipes of the if-it-was-good-enough-for-Uncle-Harry-it's-good-enough-for-me school of thought. They give the basic proportions of the huckleberry fruit to various other ingredients such as water and sugar in order to make a good, palatable wine. However, the nuances of fine winemaking are important if you wish to be even slightly scientific about the process in order to reproduce your results from year to year or just to know what's going on in that bubbling mass of fermenting berry juice. So to learn about such concerns as maintaining a constant temperature appropriate to fermentation, the difference between granulated and corn sugar, wine yeast vs. baker's yeast, acid-sugar balance, primary and secondary fermentation, measuring beginning sugar content with a hydrometer, we highly recommend a trip to your local home winemaking supply store for supplies and a few references.

Finally, you really *must* do some reading before attempting champagne, which involves fermenting wine till flat, then restarting the process in individual bottles by priming them with a little sugar. The resultant gasses given off during fermentation in the bottle make the distinctive champagne fizz, but too little or too much can be catastrophic. 'Nuf said.

Huckleberry Wine

Utensils	Use only glass, polyethylene (some plastics not suitable); possible to use unchipped enamel.
Yeast	Wine yeast is best, but if not available baker yeast.
Sugar	White granulated.
Temperature	65°-70° F. ideal.
Fermentation	Can be helped by adding yeast nutrients to help yeast reproduce.
Sterilization	It is very important to kill any foreign bacteria, from washing fruit to container to the bottling. Wine stores have tablets to put into water to aid in sterilization. Called "Campden tablets," they contain the chemical sodium metabisulphite.

6 pt. ripe berries	Crush fruit. Pour in 1 qt. of boiled water
7 pt. water	that has cooled. Mix well. Crush one
3 lb. sugar	Campden tablet in 1 T. warm water. Mix
burgundy yeast	into fruit, leave set 1 to 2 hours. Then
	take $1/_3$ of sugar, boil in 3 pts. water.
	Allow to cool to at least 100° F. Stir into
	pulp. Add yeast and ferment 7 days.

After 7 days strain pulp through fine muslin or similar material. Wring out dry. Put strained liquid into sterile container, discard pulp and boil another $1/_3$ of the sugar in 1 pt. water for one minute. Cool, then add to wine. Ferment for 10 days.

At this time transfer to another sterile container. Siphon or pour as to leave sediment on bottom. Sterilize fermenting vessel and return wine to vessel. Boil last $1/_3$ sugar in last pint water for one minute. Cool and add to wine. Ferment until bubbles stop arising to top. Liquid should be clear. If there is some residue in the bottom of the container, siphon off the wine.

Whichever type container you choose it should be covered or plugged but allow gas to escape. Plug jugs with cotton, or open top cover with plastic wrap and hold with rubber band or tie tightly.

Huckleberry Champagne

3 lb. powdered malt
1 oz. mild hops
1 gal. huckleberries
2 lb. sugar
$^1/_2$ gal. water

Combine ingredients in a large kettle. Boil 45 minutes. Add water to make 5 gallons, and sprinkle beer or champagne yeast on top. Ferment one month; should taste flat. Add 1 c. sugar and bottle. Wait one month to drink.

One resident of huckleberry country says the best part of winemaking is the bottling: "You go in to siphon, and you come out all starry-eyed."

Betty's Way

This is a way to enjoy frozen huckleberries without waiting for them to thaw. Put the berries into a small bowl, sprinkle with sugar, and add evaporated milk. Wait until the cold from the berries freezes the milk, and eat as a frozen treat.

Mildred's Way: Substitute cream for evaporated milk.

Huckleberry-Rice Salad

2 c. cooked cold brown rice
2 c. fresh huckleberries
$^1/_2$ c. unsweetened shredded coconut
$^1/_2$ c. chopped pecans or almonds
honey
$^1/_2$ c. soy milk powder
$^1/_8$ c. sea salt
$^2/_3$ c. oil (cold)
2 T. lemon juice
wheat germ

In a serving bowl, combine the rice, berries, coconut, nuts and honey ($^1/_4$ c. or to taste). In an electric blender, place the soy milk powder, salt and 1 T. honey. While blending on high speed add oil slowly until mixture becomes thick. Stir in the lemon juice. Fold soy cream into rice mixture and sprinkle with wheat germ. Yield: Four servings.

Pemmican

Mild differences of opinion exist about whether the native people of huckleberry country made pemmican. Some say they did, but most say they didn't. If you want to, here's a recipe.

1 c. dried beef or venison jerky

To make jerky, cut all the fat from beef or venison, then cut it in thin slices. Dry in oven at lowest temperature with door slightly open until meat will break and crumble. Pound the meat into powder or grind with an electric blender.

1 c. dried huckleberries
1 c. unroasted peanuts or pecans
2 t. honey
4 T. peanut butter
3-4 t. cayenne pepper

To dry berries, spread thinly on a tray and put them in the sun each day and in a warm room each night for about a week. If the weather is uncertain, they can be kept indoors all day. Or spread them on clean paper in a warm attic for about 10 days.

Add the dried berries and nuts. Here the recipe can get really elaborate with dried raisins, apricots and peaches added too, but remember to add proportionate amounts of the other ingredients for each new fruit.

Heat honey and peanut butter to soften it, then blend with the mixture. Add the pepper and make sure it's worked thoroughly through the mixture.

To go completely natural, pack the mixture in sausage casings available at meat counters—or, put the mixture in plastic bags or containers.

Put in cool, dry place and it will keep indefinitely. Pemmican can't be beat as a snack on the trail.

Huckleberry Tea

The wild huckleberry plant is used to brew a beverage like peppermint. The leaves and stems are dropped into boiling water until the desired color of tea is obtained. This is like black tea and it is very good. Steep for 3-5 minutes. Serve it hot or cold.

Huckleberry Vinegar

Add six huckleberries to a jar of vinegar to give it a delicate "blush" color, or make vinegar from huckleberry pulp. Use with wild meats.

Eating Huckleberries in Camp

Huckleberry Camp Cake

You've brought along your whole-wheat pancake mix, and you have some honey and a little Squeeze Parkay. Mix these together until it looks like a thick cake dough, then throw in as many huckleberries as you want. Mash them in if you like your cake purple (or if it'll be dark out when you eat and you don't care), or toss them in lightly to have them stay whole. Put into a well greased cooking pot—aluminum will do—and cover. Use stones or large sticks to prop the pot at a reasonable distance from some glowing coals, and proceed to spend the rest of the evening telling lies and poking the coals to get exactly the right baking action. The cake will be done just before you feel your eyes slam shut. Eat it then for some unusual dreams.

Huckleberry Slump

Heat a mess of berries in a Dutch oven over a campfire until boiling. Mix up some Bisquick, cinnamon and brown sugar and toss it in on top. Cook until done.

Field-Dressed Huckleberries

Allison's Way: Pick a big handful, open your mouth, and toss 'em back!

Chuck's Way: Carry a small bag of oatmeal along on your hike or expedition. Mix with fresh berries and enjoy as trail food.

George's Way: Collect huckleberries in a 35mm film canister until it is full. This is a perfect measure for a mouthful.

Huckleberry Patch Lunch

2 slices of bread, buttered

sugar

freshly picked huckleberries

At lunchtime, situate yourself strategically within arm's reach of a good bunch of berries. Get out your buttered bread, and fill a sandwich with freshly picked berries. Ignore any comments from jealous picking companions who did not think ahead and brought roast beef sandwiches instead.

McGarvey Family Huckleberry Dumplings

The McGarveys came up with this recipe while camping in Glacier Park in the 1930s.

1 qt. berries	Simmer in Dutch oven. Stir frequently.
1 c. water	
1 c. sugar	
Bisquick	Mix Bisquick recipe for dumplings; drop into bubbling syrup. Cook 15 minutes uncovered. Top with ice cream or whipped cream (at-home variation!).

A Note to the Reader...

Something more to say about huckleberries? Send stories, recipes, research or observations to 'Asta Bowen, $^c/_o$ American Geographic Publishing, P.O. Box 5630, Helena, Montana, 59604. Thanks for your interest.